FRONTIER AND OVERSEAS EXPEDITIONS FROM
INDIA

FRONTIER AND OVERSEAS EXPEDITIONS FROM INDIA

COMPILED IN THE INTELLIGENCE BRANCH

DIVISION OF THE CHIEF OF THE STAFF
ARMY HEAD QUARTERS

INDIA

VOL. II—Supplement A

OPERATIONS AGAINST THE ZAKKA KHEL
AFRIDIS 1908

The Naval & Military Press Ltd

Published by
The Naval & Military Press Ltd

In reprinting in facsimile from the original, any imperfections are inevitably reproduced and the quality may fall short of modern type and cartographic standards.

PREFACE.

CAPTAIN E. Hewlett, the Lancashire Fûsiliers, has compiled this official account of the Expedition, undertaken in 1908, against the Zakka Khel Afridis of the Bazar Valley. This formed the fourth occasion on which it had been found necessary to send a punitive force into this valley. During the 2nd Afghan War, both in 1878 and in 1879, columns were despatched to exact reparation for raids committed from the direction of the Bazar Valley against the line of communications through the Khaibar Pass. Towards the end of the Tirah campaign in 1897-98 an incursion into the Bazar Valley completed the visitation of the whole of the Afridi country.

W. MALLESON, *Colonel,*
Assistant Quarter Master General,
Intelligence Branch.

SIMLA;
November 1908.

CONTENTS.

CHAPTER I.

EVENTS LEADING UP TO THE EXPEDITION.

Short history of the Zakka Khel (1898 to October 1907)—Raiding gangs organised—Raids in October, November and December—Punitive measures advocated by Indian Government—Raids on Kacha Garhi Railway Station and Peshawar City—Form in which punitive measures outlined not approved—Expedition approved with restrictions—Afridi *jirga* interviewed by Sir H. Deane—Exodus from Bazar Valley—Instructions to General Officer Commanding 1

CHAPTER II.

COMMENCEMENT OF OPERATIONS TO ESTABLISHMENT OF STANDING CAMP AT WALAI.

Brief description of Bazar Valley—Plan of operations for entering the valley—Composition of the Expeditionary Force—Movements of "flying column" under Colonel Roos-Keppel—The Main Body—Advanced column under General Willcocks—Concentration near Walai—Walai Camp—Success of initial movement . . . 9

CHAPTER III.

PUNITIVE MEASURES IN BAZAR VALLEY.

Movements of the enemy—Columns formed for daily excursions—China destroyed—Reconnaissance towards Tsatsobi Pass—Halwai position attacked—Halwai destroyed—The Zakka Khel reinforcements—The Afridi *jirga* come in 14

CHAPTER IV.

OVERTURES FOR SETTLEMENT TO WITHDRAWAL OF TROOPS.

The Afridi *jirga* meet the Zakka Khel at Halwai—External interference—The Zakka Khel swear to abide by decisions of *jirga*—Return of *jirga* delayed by reinforcement of miscellaneous *lashkar*—The *jirga* return to Walai—Terms of settlement arranged and accepted—Withdrawal of troops—Details of the return march . 18

CONTENTS.

APPENDICES.

	Page
1. Despatches and Casualty List	21
2. List of Staff	29
3. Political Report	32
4. Report of Commanding Royal Engineer	39
5. Numerical Strength of Force	45
6. Signalling Report	46
7. Expenditure of Ammunition	49

MAPS.

1. Map of Bazar Valley.
2. Camp at China.
3. To accompany signalling report.

OPERATIONS
AGAINST THE
ZAKKA KHEL AFRIDIS
BY A FORCE UNDER THE COMMAND OF
MAJOR-GENERAL SIR JAMES WILLCOCKS, K.C.M.G., C.B., D.S.O.,

IN

1908

WITH A SHORT ACCOUNT OF
THE EVENTS WHICH LED UP TO THE EXPEDITION.

In the expedition of 1897-98 the forces of the British Government had effectively dispelled the cloud of mystery that had been spread over Tirah for many centuries. The country had been explored from end to end. The Afridi tribes had learnt at last, that their secluded villages and mountain fastnesses, nestling along the base of the Safed Koh, could no longer be regarded as inaccessible if the country were invaded by a determined and well-equipped force. The settlement concluded at the end of the operations proved satisfactory to both parties. The lenient and liberal terms granted by the Government were appreciated by the tribesmen whose resources were temporarily exhausted from the effects of the invasion. Recruiting in the Indian Army and Frontier Militias became as brisk as ever.

It may confidently be asserted that the stability of the peace would have been more permanent, had it not been for the existence of certain sinister influences, which continually threaten to undermine the loyalty of the border tribesmen towards the British Government. The *mullas* dwelling in tribal territory as well as in Afghanistan are, by reason of their fanatical inclinations, ever prone to stir up strife against an alien Government, while the presence of an anti-British party in Kabul, affords a sure guarantee that disaffected tribesmen can obtain encouragement and support from that centre. In the year 1904 a large body of Afridis visited Kabul. They were well received and dismissed with substantial presents of cash; they were also permitted to purchase arms and ammunition. In consequence an unusually large number of rifles, mostly Martinis, and cartridges were imported into Tirah during that year. The reception accorded in 1904 to the Afridis who went to Kabul emboldened the tribes as a whole and the Zakka Khel in particular to adopt a contumacious attitude towards the British Government. On the night of the 3rd September 1904 a daring raid on the village of Darshi Khel near Bahadur Khel was made by a band of 15 Afridis (Zakka Khel and Kambar Khel) and two Orakzais. As a consequence of this raid the well disposed among the tribal *maliks* were induced to attempt the coercion of the offending section, but with no definite result. The *maliks* confessed that the tribesmen owing to the favour shown to them at Kabul had got beyond control. The stoppage of the whole of the allowances due to the Afridis for 1904 again induced the remainder of the tribes to take concerted action against the Zakka Khel. Their endeavours produced no lasting results, but it was recognised that they had done all that could be expected to coerce the recalcitrants, and early in 1905 the allowances of all except the Zakka Khel were paid.

During the years 1904, 1905, 1906 and 1907 the Zakka Khel maintained an attitude of open defiance to Government. Bands of the tribe so harried the borders of the Kohat and Peshawar districts that security of life and property was seriously menaced. The other Afridis behaved well in so far as they restrained members of their own clans from joining in these raids, but they confessed their inability to restrain the Zakka Khel and indeed recommended to Government the occupation of the Bazar Valley, as the only effective means of dealing with the situation.

In June 1907 the Zakka Khel on finding that there was no intention of calling in their *jirga* along with other Afridi sections to receive allowances, sent a *jirga* uninvited to Landi Kotal and announced that they had come to make peace with Government, to hear the charges against the section and to settle the cases. On being informed of the claims against them they said that they were absurd and could not be entertained. They were prepared to restore any stolen property which accused men might acknowledge to be in their possession, and to return three of the six rifles taken in a raid from the Pabbi Police Station. They said that the system of deducting fines from allowances must be stopped; that their oath must be accepted in all cases; that they would accept no responsibility for raiders passing through their country to and from British territory; that in no circumstances would they surrender offenders for punishment; and that all restrictions on members of the tribe going to Kabul, or receiving allowances from the Amir must be removed. Their attitude in short showed that they had no real desire to come to a settlement, and they returned to Bazar without matters being in any way advanced.

However the uncompromising reception of this *jirga* had so excellent an effect that the borders enjoyed an immunity from their raids during the summer months. As time passed without the long expected punishment being inflicted, so the fear of punishment passed with it. In October raiding recommenced when the leaders, the chief of whom were Dadai, Usman and Multan, returned from a visit to Kabul.

The Zakka Khel who visited Kabul had been favourably received by Sardar Nasrulla Khan, those who were in receipt of annual allowances from the Amir were paid, many allowances were augmented and new ones granted. Every facility was again allowed to the tribesmen to purchase rifles, of which a large number were available from the Persian Gulf source of supply. In the mind of the Afridi the fact of receiving an allowance from the Afghan Government and joining the "Kabul party" is equivalent to pledging himself to unequivocal hostility to the British Government, so that the resumption of raiding on their return was a natural sequel to this visit to Kabul. Early in October five separate raiding gangs, whose numbers were augmented by outlaws from other tribes, were organised under the following leaders:—

1. Dadai, Anai Zakka Khel.
2. Multan ,, ,, ,,
3. Gul Baz ,, ,, ,,
4. Muhammad Afzal Ziauddin, Zakka Khel.
5. Usman Khusrogi, Zakka Khel.

These set to work at once and offences, of which the following were the most important, now began to be reported with unceasing regularity.

On 5th October a gang of some 30 Zakka Khel under Dadai raided the village of Sumari in the Kohat district, killing one Muhammadan and carrying off two Hindus and property to the value of Rs. 600.

On 28th October a gang of 30 Afridis, mainly Zakka Khel, attacked Pabbi in the Peshawar district and seized five Hindus. They fired on a party of troops at the station, seriously wounding a sepoy and a zamindar. At Tarnab they attacked and wounded the Revenue Assistant in camp and carried off his Chuprassi and horses. *En route* they robbed ten cartmen. Near Peshawar they encountered a party of police and fired on them, wounding two constables. Here they had to abandon most of their heavy booty and retreated into tribal territory.

On 13th November the village of Lachi in the Kohat district was attacked by 22 men, mainly Zakka Khel. The Post Office was looted and one villager killed and three wounded. The raiders met a party of Border Military Police and fired on them, killing two and wounding two others. They carried off four rifles, jewels and other property.

On 20th November a similar gang under Dadai fired on a village near the Bara Fort. The garrison of Border Military Police turned out and were at once attacked, two men being killed and two wounded at the first discharge.

On 24th November a large gang of Zakka Khel attacked the Marai village in Kohat, but were driven off by the troops and villagers with a loss of two men killed and three wounded, all Zakka Khel. One of the men killed was Rokhan, Karigar, Khusrogi, a well-known raider of the Dadai faction.

On 25th November Multan, with some 35 Zakka Khel, carried off 400 goats from Matanni and attacked a police post, where they were repulsed.

On 5th December the village of Masho Khel in the Peshawar district was attacked and a bania's shop looted and the owner killed by a gang of 16 persons, mainly Zakka Khel, and on the same day 12 mules working on the Khaibar Road near Jamrud were carried off to Bazar by the Zakka Khel.

These raids were not mere border affairs such as might reasonably, in the circumstances of the frontier, be passed over without serious notice. Many of them, and those the most important, were organised raids by large and well armed bodies of men on peaceful villages far in the interior of our administered districts, or direct attacks upon Government posts and property. It was evident by this time that those hostile acts of the Zakka Khel had now rendered military operations indispensable. In a despatch, dated 9th January 1908, to the

Secretary of State for India, the Government made the following proposals.

That, before having recourse to punitive methods ourselves, in order to exhaust every possible alternative, Sir H. Deane, the Chief Commissioner and Agent to the Governor General in the North-West Frontier Province, should call in a *jirga* of the Afridi clans and announce to them that unless they put an end to the present state of affairs created by the Zakka Khel, Government had decided to punish this section at once.

This action, it was pointed out, might obviate the necessity for taking punitive methods ourselves, and at the same time if the Afridis expressed their inability or unwillingness to coerce the Zakka Khel it would be tantamount to a public admission of our right to adopt coercive measures ourselves and would deprive them of any excuse for joining the Zakka Khel. The attitude of the Afridis at this time showed no disposition on the part of other sections to identify themselves with the Zakka Khel and there was little reason to anticipate that they would join in opposing our advance into Bazar.

In the same despatch (9th January 1908), it was explained that a punitive expedition on the old lines was not contemplated, but that the action proposed included the following measures—

"The recovery of a fine and the disarmament of the section. The capture and trial of the ringleaders and those implicated in the recent raids. The construction of a road which would in the future contain this section and prevent their escape into the inaccessible regions to which they now resort. The road would be maintained on lines similar to those prevailing with regard to the Khaibar, so far as they are applicable. The actual offenders in the recent raids and their supporters, if captured, would be dealt with by suitable tribunals locally assembled."

Stress was laid on the desirability of the movement on the Bazar Valley being made suddenly, and it was therefore requested that preparations might be made, so that no time should be lost between Sir H. Deane's announcement to the Afridi *jirga* and the advance of the force, should this course be found necessary after his interview with the *jirga*. Before an answer to this despatch was received the Zakka Khel perpetrated two most daring raids.

On the night of 24th January some 30 raiders attacked the Kacha Garhi Railway Station between Peshawar and

Jamrud, with the object of luring out the Border Military Police garrison and then attacking them. In this they were unsuccessful.

On the night of 28th January a still more insolent raid was undertaken. A gang of 60 to 80 men raided Peshawar City, killing one policeman, and wounding two others and two chowkidars. Property said to be worth one lakh of rupees was carried away. The Khaibar rifles spent 15 hours trying to cut them off, but owing to a telephone being out of order they were informed too late and the raiders got clear away. The situation was now so acute that immediate action was necessary.

On 31st January His Excellency the Viceroy in a telegram to the Secretary of State pointed out that it was expedient to vindicate our authority without delay. He considered that it was no longer possible to influence the position by the assistance of the *jirga*, as suggested previously in the despatch of the 9th January. He proposed that, immediately after the reasons for the expedition had been explained to the *maliks* of the other Afridi sections, the expedition should at once start for the Bazar Valley. He named two brigades with one brigade in reserve as the proposed strength of the force to be employed.

On the same day a reply was received to the despatch of the 9th January. The form in which the punitive measures were outlined in this despatch was not generally approved, as suggesting a policy of annexation or permanent occupation of the Bazar Valley; but, in view of the incessant raids, His Majesty's Government were prepared to sanction an expedition, provided that it was confined to punitive measures or blockade. His Majesty's Government also considered that the neutrality of the remaining Afridis was all that could be expected of them, and that to expect their active co-operation would be trying them too high.

On the 1st February the Indian Government still urged that, in order to prevent a recurrence of similar events in future, steps should be taken to open up the country of the Zakka Khel section. It was pointed out that a blockade on the lines of the Mahsud operations, 1901, would, owing to the geographical position of the Zakka Khel territory, be impossible without our troops entering into territory of other sections of the Afridis, whose friendship it was advisable to retain.

In a telegram received on the 3rd February from the Secretary of State, immediate action was approved of, provided that there was to be no annexation or occupation and that a strict limitation of time of duration of punitive operations was imposed. Details of the operations contemplated were asked for and it was suggested that a small detachment acting with great celerity might be sufficient for actual punitive measures in Bazar.

To this the Indian Government replied that small sallies of troops unsupported in Bazar Valley would involve great risk of disaster to our men, the result of which would undoubtedly be to set the whole frontier in a blaze. The following report of details of proposed operations was rendered—

"The Zakka Khel in the Bazar Valley can produce about 6,000 fighting men and we propose to move two brigades consisting of about a similar number, with two batteries, sappers, etc., into the valley at or near China, holding one brigade ready to support them if necessary. After occupying China the force in the Bazar Valley would act as circumstances may require with a view to the capture of the ringleaders and those implicated in the recent raids. During this period the passes surrounding the valley would be blockaded, and we propose to use a force of selected Khaibar Rifles who, we are assured, can be relied upon for the purpose, to assist in this."

These proposals were approved of by His Majesty's Government on 6th February and at the same time the following restriction was imposed on subsequent action in Bazar Valley—

"It must be clearly understood that the end in view is strictly limited to the punishment of the Zakka Khels and neither immediately nor ultimately, directly or indirectly, must there be occupation or annexation of tribal territory."

Sir H. Deane was instructed to assemble the Maliks of the Afridis and to inform them of our action against the Zakka Khel. This information was also to be conveyed to the Orakzai Maliks. As soon as the Afridi Maliks had been dismissed the expedition was to start.

In the meantime raids were still being carried out by the Zakka Khel in the Peshawar district. On the 12th February Sir H. Deane saw the *jirga* of Afridi Maliks. He explained the position fully to them and advised them to go to their tribes and assist in making the Zakka Khel come to a reasonable settlement, so that troops might be withdrawn from the Bazar Valley as quickly as possible. He dismissed the *jirga* on the same day (12th February). He reported that the attitude of the *jirga* appeared satisfactory, and that they seemed relieved to hear that punitive measures would be confined to the Zakka Khel. The Maliks returned to re-assure their tribesmen, after which they proposed to raise a *lashkar* and taking it to the Zakka Khel villages in Upper Bara to put the utmost pressure on Zakka Khel.

It was now reported that the Zakka Khel were moving their families, flocks and moveable property to Ningrahar and the Bara, and that they had buried their grain. A wholesale exodus *viâ* the Thabai Pass was also taking place. For political reasons it had been considered advisable to inform His Majesty the Amir of Afghanistan of our intention as regards the expedition. Immediately after the dismissal of the *jirga* on 12th February a *kharita* was accordingly forwarded to him in the following terms—

"I write to inform you that the Zakka Khel section of the Afridis have faithlessly broken their engagements with the Government of India and, notwithstanding the very kind and too compassionate treatment that I have meted out to them, have misunderstood my leniency and, by constant raids and murderous attacks on my law-abiding people, have filled up the cup of their iniquities. I can no longer shut my eyes to these nefarious proceedings, and I therefore write to inform you that I intend to punish these people, who deserve severe treatment, and I hope that, through the friendship that exists between us, Your Majesty will issue stringent orders to prevent any of these people from entering your territories or receiving assistance from the tribes on your side of the frontier."

Meanwhile Major-General Sir J. Willcocks, K.C. M. G., C.B., D.S.O., had been selected to command the Zakka Khel Field Force.* The force had been mobilized with the utmost secrecy. On the 12th February Major-General Sir J. Willcocks was informed that the expedition might start on the 13th February, and on the same day he moved out from Peshawar. The following instructions were issued by the Government of India :—

* With Lieutenant-Colonel G. O. Roos-Keppel, C. I. E., as Political Adviser.

"The General Officer Commanding the Expedition is vested with full political control. The Political Agent, Khaibar, will accompany him as Chief Political Officer, and as such will advise and give him every possible assistance in political matters. The authority and responsibility of the General Officer Commanding must be complete, but he shall forward to the Government of India any statement of the views of the Chief Political Officer on any question of policy affecting the attitude of any of the tribes of the Khaibar or adjacent country, if for reasons recorded the Chief Political Officer so desires.

The end in view is strictly limited to the punishment of the Zakka Khel and neither immediately nor ultimately, directly or indirectly, will there be occupation of or annexation of tribal territory.

Every possible precaution must be taken to prevent any extension of the trouble to country outside the Bazar Valley.

No terms of fine or surrender of raiders must be imposed on the Zakka Khel tribe without previous reference to the Government of India.

It is necessary to punish the persons implicated in the numerous raids which have been organised in tribal territory and which have resulted in murder and robbery in British India. For the trial on the spot when possible of any such persons who may be captured the Political Agent, Khaibar, will take action with the fullest powers conferred by Government of India's order No. 1424-F., dated 25th May 1903. Formal orders on this point follow.

All communications on political questions will be addressed by the General Officer Commanding direct to the Foreign Secretary, Government of India, and repeated to the Chief of the Staff and also to the Chief Commissioner, Peshawar. Telegrams regarding military operations will be repeated to the Foreign Secretary, Government of India."

CHAPTER II.

Commencement of Operations to establishment of Standing Camp near Walai—(13th—16th February).

Brief Description of the Bazar Valley.

The Bazar Valley, which is one of a series of parallel valleys running almost due east and west, is only about twenty miles long with a varying breadth of between eight and twelve miles from watershed to watershed. The valley itself lies at an elevation of three thousand feet. On the north the Alachi mountains separate it from the Khaibar Pass and on the south the Sur Ghar range separates it from the Bara Valley. The highest peaks of these two ranges attain elevations of from five to seven thousand feet. Through the valley runs the Bazar stream fed by tributaries draining the heights on either side, and flowing almost due east, till it joins the Khaibar stream at Jabagai. The eastern end of the valley is narrow and just before its final debouchure on to the Peshawar plain contracts into an almost impassable defile. The western end of the valley, on the other hand, is comparatively wide and open, and climbs gradually up to the magnificent snow capped range of the Safed Koh, the lower ridges of which form the boundary of the valley.

It is this upper portion of the valley which is owned by the Zakka Khel. It consists of two main branches, each about two miles broad, enclosing between them an irregular spur. This spur running out from the main watershed in a series of relatively small hills, ends in an abrupt peak, just above the great Zakka Khel stronghold, China. About two and-a-half miles east of China the two branch valleys unite and in the apex of their junction, closing the mouth of the China plain, is an isolated hill known as Khar Ghundai.

Through the circle of mountains to the south-west and west go four principal passes—the Mangal Bagh and the Bukar giving access to the Bara Valley, and the Thabai and Tsatsobi leading into Afghanistan. The two former passes may be regarded as "side-doors" giving communication to near neighbours—the clans of the Bara Valley. The Thabai and Tsatsobi passes constitute the "back-doors" of the Bazar Valley, the "bolt-holes" into Afghan territory, by which, whenever their home is threatened from the Indian frontier, the tribesmen can retire, carrying with them their women and children and all moveable property.

The great difficulty in dealing with the Zakka Khel is due to the existence of this "back-door" and the knowledge that beyond it lies a sure and safe asylum. The "front-door" to the valley is from the Khaibar Pass and over the Alachi range, across which are four passes—the Chura, Alachi, Bori and Bazar. The first three have been made use of by our troops on former expeditions. Of all four the Chura is by far the easiest and it has the advantage, or disadvantage as the case may be, of leading through the territory of another clan, the Malikdin Khel, of which the chief, Yar Muhammad Khan, professes the deepest sympathy with our cause.

General Willcocks had formed the following plan of operations for entering the Bazar Valley. To concentrate the field force at Lala China and to despatch on the day after concentration an advanced brigade

Plan of operations for entering Bazar Valley.

without transport and carrying emergency rations to China in Bazar by the Chura Pass. The rear brigade, with all transport for the force was to reach Chura on the same day and remain there ready to reinforce. Meanwhile, a flying column moving from Landi Kotal was to block the passes at the west end of the Bazar Valley.

It was, however, found that, owing to the heavy snow on the higher hills, the latter movement had to be modified, it being at that time impossible to reach the Thabai Pass on the Morga range from Landi Kotal, or to reach the Tsatsobi Pass without entering the Bazar Valley by the Bazar Pass.

On the 12th February the expeditionary force, composed as follows, was assembled at Peshawar ready to make a forward movement.

1st Brigade (Brigadier-General Anderson).

1st Battalion, Royal Warwickshire Regiment.	Sections A and B, No. 1. British Field Hospital.
53rd Sikhs.	No. 101, Native Field Hospital.
59th Scinde Rifles.	Sections A and B, No. 102 Native Field Hospital. (b)
2nd Battalion, 5th Gurkhas.(a)	Brigade Supply Column.

2nd Brigade (Major-General Barrett).

1st Battalion, Seaforth Highlanders.	Sections C and D, No. 1 British Field Hospital.
28th Punjabis	Sections C and D, No. 102 Native Field Hospital. (b)
45th Sikhs.	No. 103 Native Field Hospital.
54th Sikhs.	Brigade Supply Column.

Divisional Troops.

2 Squadrons, 19th Lancers.
2 Squadrons, 37th Lancers. (a)
23rd Sikh Pioneers. (f)
25th Punjabis.
No. 3 Mountain Battery, Royal Garrison Artillery. (a)
4 guns, 22nd (Derajat) Mountain Battery. (f)
No. 6 Company, 1st Sappers and Miners. (f)
Three Sections No. 9 Company, 2nd Sappers and Miners.(f)
No. 105 Native Field Hospital.

(a) Temporarily attached to the 2nd Brigade.
(b) Remained temporarily at the base.
(f) Temporarily attached to the 1st Brigade.

Attached.

800 KHAIBAR RIFLES.

3rd Brigade (Major-General Watkis) (in Reserve at Nowshera).

1st Battalion, Royal Munster Fusiliers. (*c*)
23rd Peshawar Mountain Battery. (*c*)
1st Battalion, 5th Gurkhas. (*e*)
1st Battalion, 6th Gurkha Rifles. (*c*)
55th Coke's Rifles.
25th Punjabis. (*d*)
Sections A and B, No. 2 British Field Hospital. (*c*)
No. 112 Native Field Hospital. (*c*)
Sections A and B., No. 113 Native Field Hospital. (*c*)
Brigade Supply Column.

Movements of the flying column.

½—2nd-5th Gurkhas, Khaibar Rifles.

Colonel Roos-Keppel was placed in command of the flying column, which was to advance from Landi Kotal. The wing of the 2nd-5th Gurkhas left Jamrud on the evening of the 12th February *en route* for Landi Kotal. On the 13th, the posts of the Khaibar Rifles, as far as Ali Musjid, were taken over by the 25th Punjabis from the 3rd Reserve Brigade and two squadrons of the 19th Lancers from the Divisional Troops. Colonel S. Biddulph was placed in command of the Line of Communications. The Khaibar Rifles on being relieved marched to Landi Kotal, at which place on the evening of the 14th, the flying column was assembled and ready for the move into the Bazar Valley.

At 4 A. M. on the 15th, Colonel Roos-Keppel left Landi Kotal with his force and by 9-15 A. M. reached the crest of the Bazar Pass without meeting with any opposition. From the Bazar Pass the force continued its march south, and arrived at China the same evening. China was found to be unoccupied and the towers and enclosures of the village gave shelter from the fire of the snipers who, as usual, opened fire as soon as it was dark. One is said to have been killed and another wounded during the night.

(*c*) Retained at Nowshera.
(*d*) Temporarily on the Line of Communication.
(*e*) Retained at Peshawar.

The main column under General Willcocks left Peshawar on the morning of the 13th February and halted that night at Jamrud. On the following day Ali Musjid was reached. On the 15th the field force left Lala China and entered the Bazar Valley by the Chura Pass. At daybreak the 2nd Brigade with certain additional troops advanced, simultaneously with the movement of the "flying column" by the Bazar Pass. General Willcocks accompanied this column. Three days' rations were carried on the person by all ranks. A echelon 1st Line Transport and detachments of No. 1 British and No. 103 Native Field Hospital only were taken. The remainder of the 2nd Brigade Baggage and supply columns marched in rear of the 1st Brigade, which followed the 2nd Brigade. The 59th Rifles from the 1st Brigade were left at Ali Musjid.

Advance of the main column.
Advanced column under General Willcocks.
2nd Brigade.
No. 3 M. B. R. G. A.
53rd Sikhs.
Wing, 2-5th Gurkhas.
Det. 6th Coy. S. & M.

As the first part of the march was through Malikdin Khel country opposition was not expected nor was it encountered. Passing Chura the column turned west up the Bazar Valley, and until Tsarkhum on the borders of Zakka Khel Country was reached, no shot was fired. Tsarkhum stands on a steep cliff overlooking the Bazar Tangi, and as the advanced guard reached this point fire was opened from the high hill Tsapara, and the left flank piquets, covered by the guns, were engaged the whole way to Walai, near where the column halted for the night. The wing of the 2-5th Gurkhas was left to occupy Tsapara and thus secure the line of communication with Chura where the 1st Brigade and baggage columns were to halt for the night of the 18th-19th. South of Walai, and in the fork formed by the junction of the Bazar and Walai nalas, stands a hill called Khar Gundai, which occupies a commanding position in the Bazar valley.

1st Brigade under General Anderson.
1st Royal Warwicks.
22nd Mountain Battery.
1 Squadron, 37th Lancers.
No. 6 Company, Sappers and Miners.
No. 9 Company, Sappers and Miners.
Baggage and Supply Columns of both Brigades.
No. 1 British Field Hospital.
No. 101 Native Field Hospital.

Contrary to all expectation this was found to be unoccupied by the enemy. Taking shelter from snipers the force bivouacked for the night in the nala about 1,500 yards due east of the crest of Khar Gundai, which was piquetted by the Seaforth

Highlanders During the night these piquets and others surrounding the camp were heavily fired on. The enemy, who appear to have been chiefly Anai Zakka Khel and Sangu Khel, were led by Dadai, the notorious raider who was severely wounded. It is said that three of the enemy were killed and seven wounded, whilst the casualties in General Willcocks' column were one killed and two wounded. Early on the morning of the 16th, signalling communication was established with Colonel Roos-Keppel's column at China and General Willcocks having selected a site for his standing camp nearer to Walai, the Landi Kotal column moved in to join him there. During the day the baggage and supply columns were brought up from Chura, where the strong fort belonging to Yar Mahomed Khan of the Malikdin Khel had been occupied by the 1st Brigade under General Anderson on the previous day.

The columns concentrate near Walai.

A piquet of the 45th Sikhs, forming part of an escort to a convoy of supplies, was attacked whilst on the lower slopes of Sara Paial hill by Anai Zakka Khel led by Multan. The hill which the enemy had occupied was carried by the 45th Sikhs with the loss of two men wounded. They were supported by the Seaforth Highlanders and mountain guns.

The site selected for the camp, a well concealed position practically in the bed of the Walai Stream, was surrounded by the Khar Gundai and a circle of hills to the north, all of which were piquetted. This position had the advantage of a well secured line of communication with the 1st Brigade at Chura, and at the same time it gave complete command of the whole valley. General Willcocks was now able to turn his attention to carrying out punitive measures. Owing to the naturally barren nature of the country it was almost impossible to inflict any serious damage upon the Zakka Khel.

Walai Camp.

The village of China, the most important settlement in the valley was the first objective, and on the following day its destruction was commenced.

CHAPTER III.

Punitive Measures in Bazar Valley (17th to 24th).

17*th February.*—Desultory firing into camp took place on the night of the 16th, but the troops had been able to provide themselves with sufficient cover and no casualties were reported. Telephonic communication had been established between Head-Quarters and the most important piquets, from which early information was obtained of the enemy's movements. During the morning the enemy were seen to be moving about amongst the hills near China and building sangars. The mountain guns opened fire from Zir Ghund and about 200 men were seen to retreat south towards Halwai. It was also reported that earlier in the morning three or four hundred had been seen leaving Jabagai, that about a hundred were in Halwai and that the towers along the Bazar *nala* were occupied. During the day most of the 1st Brigade marched in from Chura, so that sufficient troops were available to form two columns for daily operations in the valley.

18th February.—A mixed column under General Barrett moved out before day break to destroy the towers and enclosures at China. The Seaforth Highlanders and Gurkhas, moving by Sarmundo and Khwar and covered by the mountain guns, occupied the hills north of China. Lieutenant Macfadyen, attached Seaforth Highlanders, was mortally wounded in this advance. Meanwhile the remainder of the column arrived at China, destroyed the main towers and collected a quantity of wood and fodder. On leaving to return to camp the column was followed up by the enemy. The troops retired steadily. The battery and its escort in the comparatively open country south of China was specially selected by the enemy for attack. The 54th Sikhs, who were also on this flank, were hotly engaged, some of the tribesmen approaching to revolver range. By 4-30 P.M. the Gurkhas, who held the hills above China, had moved off them and the Seaforth Highlanders had also come down from the eastern slopes which they had occupied. The Zakka Khel, who made fruitless attempts to follow up the retirement, lost heavily and it was noticeable that in future they were more careful to avoid open country.

As the troops retired within the camp piquets, the enemy's fire slackened gradually and by 6 P.M. had ceased. Not a single shot was fired at the piquets during the night. The total casualties were three officers, two British rank and file, and four native rank and file wounded. It was reported on this day that large numbers of the Bazar Zakka Khel and many

Sanga Khel Shinwari volunteers were collected at the Thabai and Mangal Bagh passes, also that traders in ammunition had arrived from Afghanistan and were selling their supplies at half the usual rate. On the other hand reports were also received that *jirgas* of the other Afridi tribes were collecting with a view to coming in to arrange a settlement and that many of the tribes were anxious for peace.

19th February.—A column under the command of General Anderson marched out of camp Walai at 7-15 A.M. The object of the day's operations was to complete the destruction of China and to collect forage. No opposition was encountered during the forward movement, but the enemy endeavoured to harass the retirement. Two hundred and fifty mule loads of fodder were collected, and camp was reached at 5-30 P.M. without any casualties on our side.

20th February.—A column under General Anderson advanced towards the Tsatsobi pass in order to carry out a reconnaissance in that direction. On the way the towers at Khwar and Sarmando were destroyed. As the column approached the pass the advanced guard was fired on from the hills on both sides, and the left flanking battalions were attacked on the hills west of China. Information regarding this route, as far as the foot of the pass, was obtained for incorporation in the survey of the valley and the column returned to camp. Casualties *nil*.

21st February.—During the last few days the enemy had been seen gathering in considerable strength in the vicinity of Halwai, in the south-western corner of the valley, where they had been joined by many Sangu Khel and other Shinwaris from the direction of the Thabai pass. On the 21st two columns under Generals Barrett and Anderson moved out to deliver a combined attack on this position. Soon after daylight the 28th Punjabis occupied the heights near China and held them during the day to cover the retirement, the remainder of General Barrett's column taking the route south of China. The Khaibar Rifles and No. 6 Company Sappers moved by the Bazar nala and destroyed the towers of Kago Kamar. The enemy kept up a continuous fire on this place during the whole day. The Seaforth Highlanders on the extreme left occupied the Saran Hills and kept the enemy to the south and west of Halwai.

Meanwhile General Anderson's column moving north of China by Khwar and the Sarwakai pass debouched on to the plain north of Halwai. During this movement the enemy kept up a dropping fire from the direction of Pastakai. The two columns now advanced simultaneously. The steep cliffs overlooking Halwai were occupied without a check, the mountain guns making the enemy's sangars quite untenable. The right of General Anderson's column was well protected from any attack from the direction of the Thabai or Mangal Bagh passes by the 59th Rifles. The towers and stacks of timber in Halwai were destroyed and the force commenced its return march to camp. This was, as usual, the signal for numerous parties to come down from the Thabai direction. These opened fire at long ranges, but the Mountain Battery kept them on the move.

Both Brigades were now moving by the south of China hills, which were held by the 28th Punjabis. As the rearmost battalion arrived abreast of China, the enemy's fire had almost died away, but as the 28th Punjabis covered by the fire of the guns began to leave the hills, the tribesmen appeared in considerable numbers and closely pressed the battalion. The Punjabis withdrew steadily, but lost one man killed and eight wounded. On the left the Seaforth Highlanders and Khaibar Rifles were also attacked, the enemy advancing to close range suffered many casualties. Whilst directing his rearmost companies at this stage of the fight Major Hon'ble Forbes-Sempill, commanding the Seaforth Highlanders, was killed. The 53rd Sikhs, who formed the rear centre, covered the retirement. On the arrival of the rear guard east of China hills, the enemy drew off and only fired at long ranges. Our casualties for the day were one British officer and one sepoy killed, and ten native rank and file wounded. Not a shot was fired into camp this night.

22nd February.—The force stayed in camp and only ordinary convoy duties were done. Sangu Khel and other Shinwaris continued to arrive by the Thabai Pass, and although the most influential Maliks of the Afridi tribes were now assembling at Chura to arrange terms of peace, it was feared that the presence of these reinforcements would have the effect of delaying a settlement. On this night the tribesmen tried sniping from several directions simultaneously, and some of them again pushed up close to the piquets. A non-commissioned officer was wounded.

23rd February.—A column under General Anderson moved out during the morning to China to collect fuel, the scarcity of which was beginning to make itself felt in camp. The opposition offered was half-hearted. Two hundred and fifty mule loads of fuel were collected. During the afternoon nearly 400 men representing the united Afridi *jirga*, came in from Chura and appeared extremely anxious to arrange a settlement. The Maliks even professed themselves ready to use force to bring the Zakka Khel to terms.

On the following day the *jirga* was interviewed by Colonel Roos-Keppel. They expressed confidence in their ability to effect a settlement with the Zakka Khel and all, even the Pakkai Zakka Khel, were agreed that no settlement would be satisfactory, which did not provide for the punishment of the individual raiders. A suspension of operations for two days (25th and 26th February) was granted to allow the *jirga* to meet the Zakka Khel and discuss terms with them at Halwai.

Chapter IV.

Overtures for settlement to withdrawal of troops.

(25th February to 2nd March.)

The Afridi *jirga* left on the morning of the 25th February to meet the Zakka Khel at Halwai. They had to perform a delicate task, which was rendered even more difficult by the presence of Sangu Khel volunteers who stood round the *jirga* and, shouting abuse, endeavoured to persuade the Zakka Khel to further resistance. The *maliks* had urged that the presence of the troops in the valley was necessary in order to give weight to their representations to the Zakka Khel and that if they were withdrawn before their deliberations were completed, they could not guarantee the success of their mission. General Willcocks had meanwhile been informed that no rigid limitation was imposed on his stay in the valley, provided a settlement was quickly arrived at and followed by the early withdrawal of the troops.

The Afridi jirga meet the Zakka Khel at Halwai.

External interference.

In spite of the interference of this outside element at the *jirga*, the counsels of the *maliks* prevailed and the representative elders of the Zakka Khel took an oath on the Koran to abide by their decisions. On the 26th a water escort of the 45th Sikhs on Tsapara Hill was attacked

The Zakka Khel swear to abide by decisions of jirga.

in thick scrub jungle and lost one sepoy killed and two wounded. Early in the morning of this day terms having been arranged, the Afridi *jirga* started back to report the result to General Willcocks. However they were delayed by a miscellaneous *lashkar* consisting of Sangu Khel Shinwaris, about a thousand Ningraharis and a few Mohmands, which advanced from the direction of the Thabai Pass.

Return of jirga delayed by reinforcement of miscellaneous lashkar.

The Zakka Khel prevailed upon these to withdraw again to the western end of the valley pointing out that they had come too late to be of use and that they would only spoil the settlement which all the Afridis desired. After some demur this *lashkar* agreed to wait until the negotiations of the *jirga* were known. Owing to these delays the *jirga* arrived too late to be received at Walai on this day, and on the 27th they were met and conducted into camp by the Chief Political Officer after depositing their rifles for safe custody.

The jirga return to Walai.

The night of the 27th was spent in discussing the details of settlement, and by the afternoon on the 28th a document drawn up in Persian and giving the full terms, was presented to General Willcocks in public *jirga*. This document was endorsed by all the *maliks* and influential elders, and in it was apportioned to different clans the responsibility for the good behaviour of the various Zakka Khel sections. They promised to assist each other in punishing bad characters and agreed that Government might punish them for the misdeeds of those for whom they stood security. General Willcocks seeing that these terms more than satisfied the demands of Government formally accepted them.

Terms of settlement arranged and accepted by General Willcocks.

Meanwhile secret orders had been issued for the withdrawal of the troops on the following day. The Zakka Khel and such other of the *jirgas* as desired were allowed to proceed to China at once, and the work of strengthening the camp was carried out as usual. During this night, before the terms of agreement were known at Chura, the camp there was heavily sniped and four native soldiers were wounded.

Withdrawal of troops.

Early on the morning of the 29th General Barrett's brigade commenced their return march to Ali Musjid *viá* the Chura Pass. By 9 A.M. the baggage and transport of both brigades

Details of the return march.

had left Walai Camp and by 10 A.M. the rearmost piquets began to withdraw. General Anderson's brigade covered the withdrawal and remained at Chura on the night of the 29th, whilst General Barrett's brigade had marched straight through to Ali Musjid.

On 1st March all troops except the 59th Rifles and 2 Squadrons, 19th Lancers, which were left as a guard over the stores at Ali Musjid, arrived at Jamrud, and on the following day at Peshawar.

During the entire operations not a single follower, public or private, was killed or wounded and only one rifle was lost, belonging to a man killed whilst skirmishing in thick bush. From the time the force left Walai Camp until its arrival in British India not a shot was fired.

Results of Expedition.

The results of the expedition appeared in every way successful. The Afridi *jirga* had undertaken the punishment of the raiders and the responsibility for their future good behaviour. After the return of the troops the whole *jirga* came to Peshawar to discuss with the Political Agent the nature of the punishment to be inflicted upon the raiders. They organised a *jirga* of about 600 men, representatives of each Afridi clan, to visit the Zakka Khel settlements in turn. This *jirga* spent about a month in Zakka Khel country, living on the inhabitants according to tribal custom. This proceeding in itself constituted a punishment. They succeeded in getting hold of all the raiders except Multan, who had taken refuge in Afghan territory. They beat them and took away their booty, which was subsequently handed over to the Political Agent at Jamrud for return to the owners. They deposited some rifles as a pledge that they would not allow Multan to settle in Tirah until he had been properly punished and they further made a petition that Government would request His Majesty the Amir either to surrender him and like refugees or to expel them.

Whilst at Jamrud the *jirgas* were pestered by numerous messengers from the Mohmands and from Afghanistan urging them to join in a rising and to threaten Peshawar. Although this produced an unsettling effect no trouble resulted. The Maliks and elders of the Afridis had fully redeemed their promises and they were rewarded by Government.

APPENDIX 1.

Despatch of Major-General Sir James Willcocks.

Major-General Sir James Willcocks, after describing the operations of the force under his command, concludes his despatch as follows :—

I beg to bring to the notice of His Excellency the Commander-in-Chief the soldierly conduct of all ranks of the Force. Their good discipline and cheerfulness have been very marked, and it has indeed been an honour to command such a fine division in the field. The fact that Brigades were employed intact under their own Generals and Staff, as they had been trained in peace time, made the carrying out of all operations a simple matter, and my own share in the work was appreciably lightened by having been allowed to select my Staff Officers from the permanent Divisional establishment.

That the enemy lost heavily whilst our own casualties were small is due to the improvement in musketry training and the manner in which the troops work and run up shelter at the shortest possible notice. Sangars and pits quickly covered the bivouacs and the piquets were secured against a rush at night by wire entanglements, etc. Constant night work now forms a regular part of infantry training and the results were very plainly visible. As far as hill fighting is concerned our troops had little to learn from the Afridis.

The various departments of the force were all satisfactory and especially the transport service, where the discipline and order that prevailed were very noticeable.

The good conduct of the Khaibar Rifles, many of whom were actually serving against their own kith and kin, is a remarkable testimony to their efficiency and loyalty. Not a rifle was lost by the corps, nor was there a single desertion.

The enemy's losses, as far as can be ascertained at present, have been at least 70 killed and the wounded may reasonably be put at a much higher figure.

I have much pleasure in bringing to the favourable notice of His Excellency the Commander-in-Chief the names of the following officers and others who have rendered exceptionally good service :—

DIVISIONAL STAFF.

Brigadier-General H. Mullaly, C.B., Chief Staff Officer.

1 cannot speak too highly of this officer. From start to finish his work was done with a thoroughness which left nothing to be desired. His knowledge of Staff duties, his ability and his untiring energy in the field have all helped considerably in bringing the operations to a successful issue. I specially recommend him for advancement in the service.

Colonel A. W. Money, Royal Artillery, Assistant Adjutant and Quartermaster-General.

An excellent Staff officer who materially helped in organising the force. His past experience in the field combined with his decided ability, energy and zeal have been of the greatest assistance, and I specially commend him to the Commander-in-Chief's notice.

Captain A. B. Whatman, D.S.O., Somersetshire Light Infantry, Chief Signalling Officer.

No officer in the force did better work. The signalling to and from India and tactically in the field was of a very high order. His energy, perseverance under trying conditions, and his coolness in all circumstances are remarkable, and I strongly recommend him for advancement.

Captain N. F. C. Livingstone-Learmouth, 15th Hussars.

A fine soldier. Did very good work in the field and was most helpful in all the duties of a Staff officer.

Lieutenant A. P. Y. Langhorne, Royal Artillery, Aide-de-Camp.

An exceptionally good officer, very zealous and energetic and possesses plenty of common sense. He rendered me valuable aid in the field and I specially commend him to the Commander-in-Chief.

Captain E. T. Rich, Royal Engineers.

A very good officer, most energetic and always to the fore. His maps and reports were of great assistance during operations. He has completed a very careful survey of the Bazar Valley.

Major A. Mullaly, D.S.O., Divisional Transport Officer.

A very practical and useful officer in the field. Under his orders everything worked most satisfactorily.

FIRST BRIGADE.

Brigadier-General C. A. Anderson, C.B.

Twice commanded columns with marked success, also covered the retirement from the Bazar Valley, which operation was conducted with skill. He is a very good Brigadier; and possesses the thorough confidence of all ranks.

Lieutenant-Colonel F. M. Stewart, 2nd Battalion 5th Gurkhas.

An excellent Battalion Commander who was frequently assigned difficult duties which he invariably carried out most satisfactorily.

Captain C. de Sausmarez, D.S.O., R.A., 22nd Mountain Battery.

A gallant soldier and very good gunner, did splendidly on every occasion that the battery was employed.

Captain A. L. Tarver, 124th Baluchistan Infantry, Deputy Assistant Adjutant-General, 1st Brigade.

A good Staff Officer whose work in the field was exceedingly well performed.

SECOND BRIGADE.

Major-General A. A. Barrett, C.B.

Commanded the Brigade in the first advance into Bazar and again on 8th February near China when the enemy were very severely handled. He has much frontier experience and is a most reliable soldier.

Captain and Adjutant K. G. Buchanan, } *1st Battalion, Seaforth High-*
Major R. S. Vandeleur, } *landers.*

I specially bring these two officers to His Excellency the Commander-in-Chief's notice. The Seaforths have throughout the operations proved themselves a very fine battalion and have done a great share of the work of the force. Had Major the Hon'ble Forbes-Sempill lived I should have recommended him for a reward for his distinguished services.

Lieutenant-Colonel K. J. Buchanan, 54th Sikhs (Frontier Force).

An able and energetic commanding officer. Has a fine regiment and has shown himself a capable leader of men.

Captain H. A. H. Rice, } *54th Sikhs (Frontier Force).*
Lieutenant S. R. Shirley, }

These two officers behaved with great gallantry before China on 18th February, and I recommend them for some mark of distinction.

Captain J. P. Villiers-Stuart, 55th Rifles, Orderly Officer to General Barrett.

A good soldier, active and resourceful. Did very good work in the field.

Lieutenant C. B. Harcourt, 28th Punjabis.

Acted with coolness and much judgment on 21st February during the withdrawal from China hills.

KHAIBAR RIFLES.

Captain H. A. H. Bickford, 56th Rifles (Frontier Force).

Commanded the corps with ability and showed how well trans-border soldiers will work even against their own people when well led.

POLITICAL.

I strongly commend to the favourable notice of the Commander-in-Chief and of Government the services of—

Lieutenant-Colonel G. O. Roos-Keppel, C.I.E., Chief Political Officer with the Force.

He commanded the column which advanced from Landi Kotal on 15th February. This duty was very well carried out. He also accompanied me every day with the various punitive columns. It is due to his tact, judgment and thorough knowledge of all the Afridi tribes that the settlement of the Zakka Khel question was so rapidly and satisfactorily brought to a conclusion. I cannot speak too highly of his valuable services which are deserving of full recognition.

Khan Bahadur Sahibzada Abdul Qaiyum, Assistant to the Chief Political Officer.

I have brought his services prominently to the notice of the Foreign Department of the Government of India, and I would here only add that his assistance to the troops during our first advance was thoroughly appreciated by us all.

Mr. J. W. Littlewood, District Traffic Superintendent, North-Western Railway.

Gave me great assistance and at the shortest possible notice arranged for the many trains necessary to move up the Reserve Brigade to Nowshera.

32. For His Excellency the Commander-in-Chief's information I have attached on a separate list—Appendix A—the names of officers warrant and non-commissioned officers and men (British and Native) who did extra good work during the operations.

33. I beg to recommend the following native non-commissioned officers and men for distinguished gallantry in the field :—

28th Punjabis.

No. 2992 Havildar Hari Singh.
No. 4165 Naik Gurdit Singh.
No. 4178 Sepoy Munshi.

54th Sikhs (Frontier Force).

No. 1877 Sepoy Bishn Singh.

Khaibar Rifles.

Havildar Tar Baz.

34. A list of casualties is attached.

Appendix A.

Colonel S. F. Biddulph, 19th Lancers, Commanding the line of Communication.

Lieutenant-Colonel M. W. Kerin, R.A.M.C., Senior Medical Officer to the Force.

Lieutenant-Colonel W. J. D. Dundee, C.I.E., R.E., Commanding Royal Engineer, Force.

Major G. L. H. Sanders, Supply and Transport Corps, Chief Supply Officer with the Force.

Major C. L. Gregory, 19th Lancers, Deputy Assistant Quartermaster-General.

Major F. G. Lucas, D.S.O., 5th Gurkha Rifles.

Captain W. N. Lushington, Supply and Transport Corps, Commandant, 28th Mule Corps.

Captain P. H. Dyke, 127th Baluch Light Infantry, Commandant, 6th Mule Corps.

Captain J. R. E. Charles, D.S.O., R.E., Commanding No. 6 Company, 1st Sappers and Miners.

Captain and Adjutant C. A. Milward, 53rd Sikhs.

No. 31873 Sergeant Charlton, 3rd Mountain Battery, Royal Garrison Artillery.

Sergeant-Major Norman-Reid, The Seaforth Highlanders.

No. 2739 Colour-Sergeant John Smith, The Seaforth Highlanders.

Conductor W. J. Lyttle, 6th Mule Corps.

Native officers and non-commissioned officers.

Jemadar Mir Ahmad, Khaibar Rifles.

Subadar Mihan Singh, 28th Punjabis.

Jemadar Daud Shah, 55th (Coke's) Rifles.

Subadar-Major Amar Sing Thapa, 2-5th Gurkha Rifles.

Subadar Sayyid Ali, 53rd Sikhs (Frontier Force).

Havildar Mobin Khan, 59th Scinde Rifles (Frontier Force).

No. 698 1st class Hospital Assistant Karn Chand, 45th Sikhs.

No. 766 Driver Mangal Singh, 22nd Derajat Mountain Battery.

Return of casualties in action Bazar Valley Field Force from 15th to 29th February 1908.

SUMMARY.

Officers—1 killed, 4 wounded, *nil* missing.

Non-commissioned officers and men—2 killed, 33 wounded, *nil* missing.

Nominal return of officers killed, wounded and missing.

I.—KILLED.

Rank.	Name.	Nature of wound.
Major	The Hon'ble D. Forbes-Sempill, 1st Seaforth Highlanders.	Gunshot wound of chest penetrating heart.

II.—WOUNDED.

Rank.	Name.	Description of wound—dangerous, severe or slight.	Nature of wound.
Lieutenant	J. F. King, 3rd Mountain Battery, Royal Garrison Artillery.	Slight	Gunshot wound, left foot.
Lieutenant	P. A. F. W. A'Bekett, 3rd Mountain Battery, Royal Garrison Artillery.	Do.	Ditto right hand.
2nd-Lieutenant	Ian Campbell MacFadyen, 1st Seaforth Highlanders.	Dangerous.	Ditto abdomen (*died*).
Captain	R. M. Carter, Indian Medical Service.	Severe	Ditto left arm.

III.—MISSING—*Nil.*

Nominal return of non-commissioned officers and men killed, wounded and missing.

I.—KILLED.

Regimental No.	Rank.	Name.	Nature of wound.
		British Troops.	
9043	Private	R. Fordyce, 1st Seaforth Highlanders.	Gunshot wound of head.
		Native Troops.	
188	Sepoy	Gurdial Singh, 45th Sikhs	Gunshot wound, neck and chest with injury to arteries.

II.—WOUNDED.

Regimental No.	Rank.	Name.	Description of wound—dangerous, severe or slight.	Nature of wound.
		British Troops.		
1102	Sergeant	F. Pounds, 3rd Mountain Battery, Royal Garrison Artillery.	Slight	Gunshot wound, left ankle.
45954	Gunner	J. Simson, 3rd Mountain Battery, Royal Garrison Artillery.	Do.	Ditto left hand.
33589	Do.	E. Mitchelmore, 3rd Mountain Battery, Royal Garrison Artillery.	Do.	Ditto right hand.
22834	Do.	H. Salter, 3rd Mountain Battery, Royal Garrison Artillery.	Severe	Ditto chest.
4015	Colour-Sergeant.	C. Wright, 1st Royal Warwicks.	Do.	Ditto left thigh.
9272	Private	J. Eisthen, 1st Seaforth Highlanders.	Do.	Ditto left forearm (accidentally shot by a comrade).
		Native Troops.		
67	Driver	Mangal Singh, 22nd Mountain Battery.	Slight	Gunshot wound, right thigh.
35	Do.	Mohamed Alam, 6th Company, 1st Sappers and Miners.	Do.	Ditto of abdomen, flesh.
3084	Sepoy	Mery Karush, 9th Company, 2nd Sappers and Miners.	Severe	Fracture, left leg, caused by explosion.
2286	Do.	Arik Sawami, 9th Company, 2nd Sappers and Miners.	Do.	Contusion, right shoulder by explosion.
4557	Do.	Bhag Singh, 23rd Pioneers	Do.	Gunshot wound, left leg.
4940	Do.	Khem Singh Do.	Do.	Do. right hand.
3986	Do.	Imam Din Do.	Slight	Do. left hand.
4319	Do.	Khaja Mohamed, 28th Punjabis.	Dangerous	Do. abdomen (*died*).
4601	Do.	Abbas Khan, 28th Punjabis	Severe	Gunshot wound of neck.
4686	Do.	Kapura Do.	Do.	Do. right thigh.
2821	Naik	Kapura Do.	Do.	Do. right forearm.
2824	Colour Havildar.	Abdulla Khan Do.	Do.	Do. right thumb.
3592	Sepoy	Gurmukh Singh Do.	Slight	Do. left shoulder
4244	Do.	Rama Do.	Do.	Do. left hand.

I.—WOUNDED—contd.

Regimental No.	Rank.	Name.	Description of wound—dangerous, severe or slight.	Nature of wound.
		Native Troops.—continued.		
3954	Sepoy	Revalu 28th Punjabis	Dangerous	Gunshot wound, abdomen (*died*).
4960	Do.	Sheama Do.	Severe	Gunshot wound of testicle and right thigh.
..	Subadar	Sangat Singh, 45th Sikhs	Do.	Gunshot wound, right thigh.
4680	Naik	Mal Singh Do.	Do.	Do.
253	Sepoy	Karm Singh Do.	Do.	Gunshot wound, right forearm.
4226	Do.	Ishar Singh Do.	Slight	Do.
365	Do.	Bhagal Singh Do.	Severe	Gunshot wound, left hand.
4906	Do.	Suhail Singh Do.	Do.	Do. right leg.
265	Do.	Massa Singh Do.	Do.	Do. left forearm.
297	Do.	Santa Singh, Do	Do	Do left forearm.
2003	Lance Havildar.	Partap Singh, 54th Sikhs	Do.	Gunshot wound, both thighs.
3247	Sepoy	Prithi Singh, 59th Rifles	Do.	Do. right side, back.
2121	Rifleman	Babram Singh Thapa, 2-5th Gurkhas.	Do.	Gunshot wound, left forearm.

III.—MISSING.—*Nil*.

Nominal return of followers killed, wounded or missing.

I.—KILLED.—*Nil.*

II.—WOUNDED.

Rank.	Name.	Description of wound—dangerous, severe, or slight.	Nature of wound.
Bhisti	Lachhmia, 9th Company, 2nd Sappers and Miners.	Slight	Contusion, back, caused by explosion.

III.—MISSING.—*Nil.*

APPENDIX 2.

Detail of Staff.

General Officer Commanding the Force.	Major-General Sir James Willcocks, K.C.M.G., C.B., D.S.O.
Aides-de-Camp.	Captain N. J. C. Livingstone Learmonth, 15th Hussars.
	Lieutenant A. P. Y. Langhorne, R.A.
Chief Staff Officer.	Brigadier-General H. Mullaly, C.B.
Assistant Adjutant and Quartermaster-general.	Colonel A. W. Money.
Deputy Assistant Adjutant-General.	Captain A. W. Peck, 22nd Cavalry.
Deputy Assistant Quartermaster-General.	Major C. L. Gregory, 19th Lancers.
Deputy Assistant Quartermaster for Intelligence	Captain J. Campbell, Argyll and Sutherland Highlanders.
Intelligence Officer	Captain S. F. Muspratt, 12th Cavalry.

Attached.

Commanding Royal Engineer	Lieutenant-Colonel W. J. Dundee, C.I.E., R.E.
Senior Medical Officer	Lieutenant-Colonel M. W. Kerin, R.A.M.C.
Senior Veterinary Officer	Lieutenant A. J. Thompson, A.V.C.
Divisional Supply Officer	Colonel C. V. W. Williamson, Supply and Transport Corps.
Divisional Transport Officer	Major A. Mullaly, D.S.O., Supply and Transport Corps.
Signalling Officer	Captain A. B. Whatman, D.S.O., Somerset Light Infantry.
Provost Marshal and Treasure Chest Officer.	Captain P. Howell, Queen's Own Corps of Guides.
Survey Officer	Captain E. T. Rich, R.E.

1st Brigade Staff.

General Officer Commanding . .	Brigadier-General C. A. Anderson, C.B.
Orderly Officers . .	{ Lieutenant L. Forbes, 57th Rifles. Lieutenant H. F. Elgee, South Wales Borderers.
Deputy Assistant Adjutant-General	Captain A. L. Tarver, 124th Infantry.
Deputy Assistant Quartermaster-General.	Captain E. E. Barwell, 57th Rifles.

Attached.

Brigade Signalling Officer . .	Major A. F. Fergusson Davie, C.I.E., D.S.O., 53rd Sikhs.
Brigade Supply Officer . .	Major M. R. de B. James, A.S.C.

2nd Brigade Staff.

General Officer Commanding. .	Major-General A. A. Barrett, C.B.
Orderly Officer	Lieutenant J. P. Villiers-Stuart, 55th Rifles.
Deputy Assistant Adjutant-General.	Major H. M. Allen, 25th Cavalry.
Deputy Assistant Quartermaster-General.	Captain H. H. Norman, Northampton Regiment.

Attached.

Brigade Signalling Officer . .	Lieutenant D. K. McLeod, Queen's Own Corps of Guides.
Brigade Supply Officer . .	Captain W. C. W. Harrison, Supply and Transport Corps.

3rd (Reserve) Brigade Staff.

General Officer Commanding .	Major-General H. B. B. Watkis.
Orderly Officer	Captain H. A. Holdich, 1st Battalion 5th Gurkhas.
Brigade-Major	Major G. B. H. Rice, 31st Punjabis.

Attached.

Brigade Supply Officer . .	Captain A. R. B. Shuttleworth, Supply and Transport Corps.

Line of Communication Staff.

Commanding Line of Communication.	Brevet-Colonel S. F. Biddulph, 19th Lancers.
Staff Officer, Line of Communication.	Lieutenant R. D. C. McLeod, 19th Lancers.

Attached.

Supply Officer	Major G. L. H. Sanders, Supply and Transport Corps.
Transport Officer	Major G. A. Stewart, 7th Rajputs.
Signalling Officers	Lieutenant R. E. Barrow, 38th Dogras.
	Lieutenant D. P. Sandeman, Queen's Own Corps of the Guides.
	Lieutenant W. Gibson, 1st Battalion, Northumberland Fusiliers.
Officer Commanding Small Arms Ammunition Section, Ammunition Column.	Lieutenant A. P. Wavell, Royal Highlanders.

Base Staff.

Base Commandant	Lieutenant-Colonel A. R. Dick, 22nd Cavalry.
Deputy Assistant Adjutant and Quartermaster-General at the Base.	Major H. R. Blore, King's Royal Rifle Corps.

Attached.

Senior Medical Officer	Lieutenant-Colonel H. H. Brown, M.B., R.A.M.C.
Ordnance Officer	Lieutenant I. A. Finnis, R.A.
Base Supply Officer	Major H. L. D. Fordyce, Supply and Transport Corps.
Base Transport Officer	Major W. L. R. Amesbury, Supply and Transport Corps.
Railway Staff Officer	Major L. A. S. Hanmer, 21st Cavalry.
Commandant, British Troops Depôt.	Lieutenant H. C. Sinnott, Royal Warwickshire Regiment.
Commandant, Native Troops Depôt.	Captain C. J. White, 53rd Sikhs.

APPENDIX 3.

Political Report of the Bazar Valley Expedition, 1908.

It is not necessary, for the purposes of this report, to recount in detail the events which led to the expedition. Suffice it to say that bands of Zakka Khel had for the past three or four years so harried the borders of the Kohat and Peshawar districts that all security of life and property was lost.

The other Afridis had behaved well in so far that they had restrained members of their own clans from joining in these raids, but they had made only half-hearted efforts to check the Zakka Khel from raiding and appeared to consider the matter no concern of theirs.

It became at last absolutely necessary to punish the Zakkas, but the Government of India were naturally anxious that the punishment should be confined to them, and that, as far as possible, measures should be taken to isolate them and to avoid a general rising, which is always possible when troops move into any tribal territory. With this object the *jirgas* of all the Afridi tribes, excepting the Zakka Khel of Bazar, were summoned to Peshawar, and it was explained to them that, as Government had no quarrel with them and had no desire to extend its territories at their expense, it behoved them in their own interests to refrain from joining the Zakka Khel, who were about to be punished, to do their best to bring the latter to reason and to obtain from them such guarantees as would ensure their future good behaviour.

Simultaneously with the dismissal of the *jirgas* the Bazar Field Force left Peshawar. The Zakka Khel expected that an expedition would be sent against them, but they counted on some delay and on slow movement on our part. In consequence they deferred the removal to a safe place of their families, flocks, and portable property, until they got news of the actual start of the force which once started, moved with such rapidity as to reach Walai, a point which commands the entire valley of Bazar, on the third day after the *jirgas* left Peshawar, while the majority of the Zakka Khel were still arranging for accommodation for their families in Afghanistan.

The duration of the expedition was rigidly limited by the orders of Government and the problem how, in a very short time, to so punish the Zakka Khel as to induce them to make any submission or to give any guarantees for the future, appeared almost insoluble. Fortunately, we had to deal with a very gallant enemy, who assisted the solution by fighting in so determined a manner on the 18th, 19th, 20th, and 21st February as to suffer very heavy loss. In spite however of the resistance of the Zakka Khel, their "castellos" or small forts were steadily demolished and large supplies of wood and fodder were obtained from them. The enemy's casualties in the fighting between the 15th and 21st February exceeded those of all the Afridis in the Tirah Campaign of 1897-98. These casualties, their material losses, the hardships suffered by their families, who were enjoying the grudging hospitality of their Afghan neighbours, and their total inability to inflict serious loss upon our force, had their effect upon the Zakka Khel who, though they were too proud to sue for peace, became disposed to welcome the intervention of the other Afridis, which they would have scouted earlier.

While the fighting had been going on in Bazar, the Chiefs and elders of the other Afridi clans had been exerting themselves to the utmost to restrain the turbulent element in Tirah from joining the Zakka Khel and they were so successful that of the whole tribe probably not more than fifty volunteers from the Afridis took part against us. When they were satisfied that they could control their own tribesmen, the Chiefs and elders entered into negotiations with the Zakka Khel and their preliminary overtures were received at least with courtesy. Unfortunately, there exist both in India and in Afghanistan mischief-makers who are never so happy as when setting others by the ears and the Zakka Khel were in constant receipt of messages from Peshawar encouraging them to hold out as Government had ordered the withdrawal of the force, and from Afghanistan promising them material assistance. These messages, taken with the fact that numbers of Afghan volunteers and large supplies of ammunition and flour were arriving daily at the Thabai Pass from Afghanistan, undoubtedly stiffened the backs of the Zakka Khel, and up to the 25th February the issue of negotiations appeared very doubtful. On that date the Airidi representatives who had assembled in Sir James Willcocks' camp left for Halwai. There they held a consultation with the Zakka Khel which lasted that day and night. Their discussions were continually interrupted by the howls of hundreds of Afghan fanatics, mullas and others, who clustered on the hills around, abusing the *jirga* and calling upon the Zakka Khel not to listen to "the *feringhi* dogs". In spite of this a working arrangement was come to by which the Zakka Khel elders placed themselves unreservedly in the hands of the Afridi representatives, details being left for settlement later, and it was decided that the combined *jirgas* should proceed to China next day, invite the Political Officer to meet them there, and settle the matter.

On the morning of the 26th, when the *jirgas* were about to start, it was seen that a large force was advancing east from the Thabai Pass. It consisted of Afghans (mostly Shinwaris) with standards, who were headed by bands of mullas including some of the most influential in the Ningrahar province. The Afridi Chiefs and elders believed that they had been deceived and accused the Zakka Khel of treachery, but the latter swore that they had no knowledge of this reinforcement and that they would do their best to induce it to return to Afghanistan. The Zakka Khel elders went off to meet the Afghan force and as the other Afridis did not place much confidence in them, four elders from each of the other clans accompanied them to guard against any double dealing. The Zakka Khel informed the Afghans that they would have been delighted to see them ten days earlier, but that now that they had decided to make peace their presence was inconvenient. The Afghans were very loth to return, but the Afridis swore that if they did not do so they themselves would attack them in flank. At last it was arranged that the Afghans should return to their own border and there await the result of the negotiations. If these proved unsuccessful their assistance would be welcomed.

All this took time and the *jirga* arrived at China in the evening instead of the morning of the 26th February. They spent the night there, and next day the whole assembly, numbering about eleven hundred men and including about three hundred Zakka Khel, marched into camp. Most of them were armed, and it was necessary for the safety of the camp to disarm them. This was a very delicate task, but it was eventually accomplished. The *jirga* discussed with the Political Officer and the Assistant Political Officer the whole question of a settlement throughout that afternoon and evening and the following morning (28th February), and at last a draft agreement was arrived at which ran as follows:—

"We, the Maliks and elders of the Afridi tribe in *jirga* assembled, humbly represent to the British Government that, being anxious to end the quarrel between the Zakka Khel Afridis and the British Government, which has been caused by the misconduct of the former, we have agreed and do promise:—

<small>Vernacular copy attached marked (A).</small>

"That we, the Afridis, will hold ourselves responsible jointly and severally for the future good behaviour of every section of the Zakka Khel tribe; thus—

"The Malikdin Khel Afridis will be responsible for the Sahib Khel half of the Anai, Paindai, and Jamal Khel Khusrogi Zakka Khel.

"The Kambar Khel Afridis will be responsible for the Shan Khel Zakka Khel of Bara and Tirah.

"The Kuki Khel and Nikki Khel Pakhai Afridis will be responsible for the Mohit Khel half of the Anai Zakka Khel.

"The Sipah Afridis will be responsible for the Ziauddin Zakka Khel.

"The Kamarai Afridis will be responsible for the Khusrogi Zakka Khel, excepting the Jamal Khel.

"The Khyber Pakhai Zakka Khel will be responsible for the Bara Pakhai Zakka Khel.

"Also we promise that when we are called upon we will assist each other in punishing these bad characters, and Government may punish us by fine, by exclusion from British territory, or in any other way for the misdeeds of the Zakka Khel sections for whom we stand surety.

"As regards the past we beg that Government in its strength will take into consideration the losses suffered by the Zakka Khel, by exclusion from British territory and by war, and will not complete the ruin of a tribe for the sins of the badly-behaved minority.

"As regards the actual thieves who have been leaders in the raids into British territory we also beg that we, with the assistance of the Zakka Khel elders, may undertake their punishment wherever they may be, and we will punish them to the satisfaction of Government and in earnest of this we here deposit fifty-three rifles which are worth, according to the prices current in our country, more than twenty thousand rupees, as security, and these rifles will only be returned to us when the Political Agent, Khyber, is satisfied that the thieves have been sufficiently punished, and we, the Zakka Khel, agree to this petition and promise to assist the *jirga* in every way.

"Further we hope for the mercy and favour of Government."

This draft was approved by Sir James Willcocks, and arrangements were made for him to receive the *jirga* at 3-30 P.M., on the 28th February.

The rifles mentioned in the petition, which were to be deposited by the *jirga*, had to be obtained from them and this was a tedious business, as each person was most anxious that his neighbour's, not his own, rifle should be given up. Also efforts were made to induce us to accept arms of local manufacture which have a very small market value. These difficulties were at last got over, and Sir James Willcocks received the representatives of the whole tribe, including the Zakka Khel. The petition of the *jirga* was read to them in Persian and explained in Pushtu and at the termination of the reading each Chief rose and formally asked the elders of his clan whether they authorised him to guarantee these terms on their behalf and on behalf of their constituents. The reply being unanimous, Sir James Willcocks accepted and signed the settlement on behalf of Government. When the settlement had been signed, the elders sent messengers to every part of Bazar to inform the Zakka Khel that peace had been made and that they could now bring back their families to their houses. They also warned them that if they discredited the settlement by firing at the troops the Afridi *jirga* would attack them. As the weather was very bad some of the Zakka Khel members of the *jirga*, whose homes were near, asked to be allowed to depart and were permitted to do so. They sent some of their men to Thabai to dismiss the Afghan gathering.

Next morning, the 29th February, the whole force marched to Chura. All the *jirga* accompanied the troops except the Zakka Khel elders, who waited until the rear-guard had left camp and then took leave of us in the most friendly manner.

Not a single shot was fired at the troops during the march of 29th February to Chura, during the night there or during the march of the following day to Ali Masjid, a record in the history of frontier expeditions.*

The campaign has been a most valuable and interesting test of the extent of our control over our border tribes. That the Chiefs and elders of the other Afridi clans should have succeeded in restraining their clansmen from joining the Zakka Khel, with whom they are so intimately connected, is little short of marvellous, and that they should have bound the Zakka Khel in an agreement by which they themselves accept a very onerous responsibility is hardly less so. The Afridi Chiefs and elders have worked hard and well, they have shown determination and unselfishness and have deserved well of Government. Among so many it is difficult to choose, but I cannot refrain from mentioning the names of Malik Zaman Khan, Chief of the Kuki Khel, Malik Yar Muhammad Khan (Malikdin Khel), Malik Shah-Mard Khan (Sipah), and Malik Amaldin (Kambar Khel). Of these the first did most; in fact without him I doubt whether a settlement could have been arrived at. I am making a separate representation on his behalf. Malik Yar Muhammad Khan was originally, it must be admitted, responsible to some extent for the Zakka Khel trouble, as his intimacy with some of the raiding leaders and his intrigues with Kabul did much to encourage the Zakka Khel in their misconduct. He has, however, worked so well and suffered so much inconvenience throughout this expedition that he may be considered to have purged his past offences. He gave up without demur his fort at Chura to be garrisoned by our troops, although this violation of his privacy was most distasteful to him, and he did not spare himself throughout the tedious and difficult negotiations of the *jirgas*.

Another point of great political interest was the behaviour of the Khaibar Rifles. This corps, which is mainly composed of Afridis, including some 350 Zakka Khel, had to take part in an expedition against a people to whom the men were bound not only by race and religion, but by the closest ties of blood. Indeed in many cases during the expedition brother was fighting against brother and son against father. The experiment was viewed with mistrust by many and with misgiving by all, but it has been more than justified, as throughout the expedition the Khaibar Rifles gained universal praise for their keenness, and willingness, not a man deserted, and not a rifle was lost. I mention this particularly in this report, as I consider it of great political importance. The Khaibar Rifles were in 1899 merely an armed rabble without cohesion or discipline and the tribesmen themselves have fully shared the doubts of the authorities as to whether the men could really be employed against their own relatives and countrymen. The political effect of the satisfactory behaviour of the corps cannot but be considerable, not only among the Afridis, but throughout the border tribes.

---

* Not accurate. After peace had been declared in April 1898 at the close of the Tirah expedition the withdrawal from the Bara Valley was unmolested. Other instances of peaceful withdrawals after operations in Waziristan, Orakzai country, etc., can also be quoted.

I have made every effort to obtain an accurate estimate of the losses of the enemy, but it is most difficult to get anything like complete figures. I shall know later, but at present I can only say that I know the names of 38 men buried in Bazar between the 15th and 21st February. A good many corpses were taken to Afghanistan for burial and some to Tirah. Also we have to add the enemy's losses on the 21st February, when they are believed to have been heavy, and on the 23rd February, when they are believed to have been slight. Altogether I think that it is safe to say that the enemy have lost not less than 70 killed. As regards the wounded, Afridis are very disinclined to give details of their losses. This is partly owing to pride and partly owing to fear of the evil eye, to which wounded men are particularly susceptible. We shall probably never know accurate details of the enemy's wounded and can only estimate their losses by the usual proportion of wounded to killed which would bring their total casualties to a very high figure. But for these losses it would have been impossible to bring the tribe to reason without a protracted occupation of the country, and it is only the remarkable military success of the expedition which made a settlement feasible.

The Afridis, who are no mean judges of hill fighting, express themselves amazed at the handling and conduct of the troops as unlike anything they have seen or heard of, and the fact that they have obtained no loot in mules, rifles, stores, or ammunition, on which they confidently counted to compensate them for their own losses, has given them a strong distaste for expeditions conducted on these novel lines.

As regards the punishment inflicted upon the enemy by material loss, every fort and watch tower in the valley has been razed to the ground, large supplies of wood and fodder have been obtained and the losses due to exposure among the sheep, goats, and cattle (which are always, in Bazar, kept in caves in the winter) have been very great. The Zakka Khel have entirely exhausted their supplies of ammunition and cash and every individual in the tribe has felt the sharp punishment inflicted and will continue to do so for some time. I think that there is no doubt that the Zakka Khel really repent their misconduct, and that they will do their best to prevent further trouble. Dadai, the cleverest and strongest man in the tribe, who was dangerously wounded early in the expedition, sent a message to his clansmen strongly urging them to make peace and saying how much he repented his share in bringing punishment upon them. He said that he had been made a catspaw by Sardar Nasrulla Khan, who had given him pay for forty men to raid into British territory, had assured him that the Government of India would never send an expedition to Bazar, and had promised him protection and assistance for the whole tribe. Dadai urged his fellow tribesmen to make peace with the English and to have no more to do with Kabul. The Zakka Khel trouble had its origin in Afghan intrigue and the same influence was again employed to prevent a settlement, but without success. It appears certain that His Majesty the Amir issued a proclamation forbidding his subjects to help the Zakka Khel or to take part in the war, but it is no less certain that this proclamation was openly disregarded, not only by His Majesty's subjects but by his own officials, and this at a time when the Amir himself was at Jalalabad close to the scene of action.

In spite, however, of this and, thanks to a combination of military skill, political good fortune, and the loyal efforts of the Afridi Chiefs, a settlement of the Zakka Khel question has been arrived at which is more satisfactory than I should have dared to hope for even had our movements been unhampered and the duration of our stay unlimited, and I believe that the settlement will prove a durable one and that it will improve our relations not only with the Zakka Khel, but with all the Afridis. Should Government be satisfied with the political results attained, and should my hopes for the future be justified by the event, I would like to state that such political credit as may have been gained is due mainly to the efforts of Sahibzada Abdul Qaiyum, whose services have been invaluable. The Sahibzada served through the Tirah Expedition and has been since 1898 Assistant Political Officer, Khaibar. His knowledge of the Afridis and the real affection and respect which they have for him have given him a power over them which has ever been used for their good. The recent negotiations were complicated and difficult, and we had many a *mauvais quart d'heure* when it looked as if all would be broken off.* We were hampered by the necessity for haste, and intriguers from both the British and Afghan borders worked against us, but throughout the negotiations Sahibzada Abdul Qaiyum never despaired; nothing could ruffle his imperturbable patience or bend his determination, and that we have made a real settlement is due principally to his influence and skill. I would take this opportunity of bringing his services strongly to notice.

In conclusion, I venture to express to Sir James Willcocks my very real gratitude for his kindness, advice and support, without which his political staff could have done little to serve him.

G. O. ROOS-KEPPEL, *Lieut.-Colonel,*
Chief Political Officer, Bazar Valley Field Force.

APPENDIX 4.

Engineer Report by Lieutenant-Colonel W. J. D. Dundee, C.I.E., R.E., Commanding Royal Engineer, Bazar Valley Field Force.

Preliminary.—The operations of this force in advance of Peshawar lasted from 12th February 1908 to 2nd March 1908 and this report deals with this period and the preparations prior to the same.

These preparations were commenced by me on 4th February 1908, but as the advance was kept secret the more conspicuous preparations had to be postponed till 12th February 1908. The troops available for engineer operations were No. 6 Company, 1st Sappers and Miners, and No. 9 Company, 2nd Sappers and Miners, and 23rd Sikh Pioneers. Of the latter however, one wing was permanently stationed at the Chura Kandao and employed solely on piquetting duties.

1. *Defence Works.*—Camp perimeter entrenchments were entirely carried out by the troops and except in the case of the camp at Chura, were very well done. I undertook the supervision of these entrenchments myself and strengthened the Chura camp defences on 29th February 1908 on arrival from Walai. The entrenchments consisted as a rule of earth parapets for use kneeling or lying down with the ground in rear cut out for the men to sleep in. Breastworks of dry stone walling were given on hard stony sites and head cover was liberally supplied in the form of stone loop-holes, boulders placed at intervals of 2 feet or so along the crest of the parapets, and sprigs of privet stuck into the crest. Traverses were given at 5 yards or 10 yards intervals according to site. Piquets were always sangared with breast works of dry stone walling. I visited nearly all the piquets in the vicinity of Walai and supervised the strengthening or rebuilding of the sangar walls. The remainder of the piquets were visited and strengthened under the orders of Royal Engineer officers and four piquets on Seaforth Hill (Khar Gundai) were rebuilt by parties of No. 9 Company, 2nd Sappers and Miners.

Piquets being of course more exposed to attack than the perimeter, the supervision of Royal Engineer officers was largely concentrated on to the defences of the piquets which were much stronger than similar defences in Tirah in 1897.

(b) *Obstacles.*—Abbatis of thorn bushes was liberally used round piquets, and empty beef, etc., tins were placed in the abbatis. Abbatis of trees was used to close the river bed above and below the Walai camp. Barbed wire was put round the outlying piquets of the Walai camp, either laid on the ground or raised by means of stone pillars or iron posts. The only barbed wire fences given were round an inlying piquet south-east of Walai, along part of the perimeter of the same camp and round the supply godowns at Jamrud. Dangerous gaps between piquets were protected by barbed wire.

(c) *Fougasses.*—Explosives had to be economised and reserved as much as possible for the destruction of towers and village defences. Fougasses were therefore only laid at spots which I noticed the enemy frequented. The first fougasse was laid on one of the China summits on 23rd February 1908, it was timed to explode in 15 minutes (30^1 fuze) after the last men of the piquet had rushed from their sangar on retirement and fired by Lieutenant Campbell, R.E., who left the sangar with the last eight Gurkhas. The enemy on that date followed up the piquet more slowly than previously, and in consequence the fougasse went off before the Zakkas were close to it. No one is said to have been killed by it, but so much surprise was caused that the enemy delayed reoccupying the hill and did not press the rear guard, as on other days, from that flank at all.

The 2nd fougasse was laid 150 yards outside the piquet on Gurkha hill. It was laid on 23rd February 1908, but the enemy saw the work being done, and kept clear of that hill till the night of the 24-25th February, when they came up to attack the piquet. Lieutenant Bassett, R.E., exploded the fougasse electrically from the piquet, firing stopped entirely for 2 or 3 minutes, was then resumed for a short time and then stopped entirely. It is said that four corpses were carried off from this hill and the moral effect of these explosions was considerable. In each case the charge was 30 pounds of gun-cotton.

I think fougasses may well be used in trans-frontier warfare.

As economy of transport limits the amount of explosives that can be carried, and, as a Royal Engineer officer is required for the firing of each fougasse, their employment should be limited to such spots as it is known an enemy will visit.

In a retirement fougasses may well be placed near a village or piquet which has to be evacuated. Firing must be done by fuze, unless one is prepared to abandon with each fougasse about half a mile of cable, as the officer commanding the rear guard cannot as a rule spare the time to re-advance in order to recover the line. Outside piquets, which the enemy snipe habitually, the fougasse electrically exploded may well be used.

(d) *Hand Grenades.*—Hand grenades were not used. On steep ground I always order a heap of boulders to be collected inside a sangar. When the enemy comes close up boulders can be thrown over the parapet without a man being exposed, as he is if he fires his rifle. The Afridis in such cases frequently throw rocks from behind cover, when at close quarters, and in this expedition the Seaforth

Highlanders had one officer and five or six men in one piquet all more or less severely hit by rocks. In such a situation I think a percussion hand grenade would be most useful and I recommend experiments with the "Improved Shrapnel Hand Grenade" which is made by "The Cotton Powder Company, Ltd., 32 Queen Victoria Street, London, E. C." If satisfactory a few men per regiment could be trained and employed at night in exposed piquets.

2. *Demolitions.*—These consisted of the destruction of towers and the defences—loop-holed walls—of fortified villages.

The towers destroyed were :—

Three in village just east of Walai.

Two in the Walai camp.

The towers at Jabagai, Halwai, Khwar and Sarmundo.

Multan's tower and thirteen towers in China.

When time was not very restricted the usual rule of one pound of explosives per foot run of diameter of a circular tower, or of one side of a square tower 50 per cent. was followed.

Gun-cotton was always used, dynamite is not suitable.

A solid base tower has horizontal layers of brushood every three or four feet. A tunnel should be driven in the centre of the tower or as near as possible just under one of these layers and the whole charge placed there. A layer four or five feet above the ground is at a convenient height.

In a hollow tower the charge should be divided into three parts and placed inside the walls at floor level in three corners. The time required is about one-and-a-half hours for a solid tower and half an hour for a hollow tower. If this time is not available it is quicker in the case of a solid tower to sink a hole three or four feet deep in the centre of the first floor and place the charge there. This would take about half an hour, but it is not so effective.

A hollow tower can be demolished more quickly by using more charges with little if any tamping.

The method of blowing up must depend upon the time the officer commanding the force can allow, if time be sufficient the usual economical method is followed.

Loop-holed walls were usually pulled down. In many cases they were first undermined with a pickaxe and an occasional slab of guncotton was used.

3. *Bivouacs.*—I have no remarks on this subject, the troops dug themselves in and rigged up waterproof sheets and stone walls. Officers usually had *tentes d'abris.*

4. *Water-supply.*—The water-supply arrangements excluding those in paragraph 15, consisted merely in marking, drinking, watering, and washing places and in canalizing the stream at Walai.

Water was particularly good in the Baza Valley, analysis showing nothing objectionble except an excess of hardness which is always the case in a limestone country.

5. *Drainage.*—The only drainage work done was a surface drain run through the 2nd Brigade Field Hospital camp at Walai and thence east past the Seaforth camp.

6. *Sanitation.*—Two incinerators were erected at Walai.

7. *Camp Communications.*—The camp communications were of the usual type. Three egress roads from the railway sidings to the supply depôt camp were made at Jamrud. Ramping was done at Lala China and Ali Musjid. Good tracks were made in and out of the Chura camp. Numerous ramps and roads were made inside the Walai camp, which was by far the least flat of all the camps. Paths for pakal mules were made up the 45th Sikh hill and the Seaforth hill (Khar Ghundai).

8. *Road Work.*—In February 1898 a camel road had been made from the Khaibar stream half a mile below Ali Musjid to a quarter mile on the far side of the Chura Kandao. This road was thoroughly repaired, improved and extended to the Chura camp between which and the Khaibar there is now an excellent camel road with a very good surface, minimum width 8 feet and ruling gradient 1 in 8.

9. *Nala Crossings.*—No work was required.

10. *Bridges.*—There were no bridges.

11. *Stores and Explosives.*—The Sapper and Miner Companies and battalion of Pioneers had their equipment and stores as authorised.

In addition, in the first advance were sent to the front:—

       5 mule loads of posts for barbed wire fencing.
      10 mule loads of barbed wire.
       5 mule loads of sand bags.

At Ali Masjid an advanced depôt of No 1 Engineer Field Park was established containing:—

	Mule loads.
2 anvils ¾ cwt.	1
Picks with helves	10
Spare helves for picks	5
Shovels	5
Barbed wire	10
Uprights and piquets for barbed wire	10
Jumpers	3
Sandbags	1

	Mule loads.
30 pounds binding wire	
10 ,, wire nails 1½″	
30 ,, ,, 2½″	½
16 pliers side cutting	
6 pairs gloves	
12 bamboo poles for barbed wire	
200 pounds dynamite	2
1,570 pounds weight guncotton	15
1,500 detonators, commercial for guncotton	1
350 ,, Service No. 8	
10 boxes dry guncotton	1
4,800 feet of safety fuse	2
4 tarpaulins	
6 hammers, smith	½
48 helves for felling axes	

From this advance depôt the following stores were sent up to the front :—

690 pounds of weight guncotton	7
50 detonators, Service No. 8	by hand
1,200 feet of safety fuse	
36 spare helves for picks	1
12 helves for felling axes	

All stores were packed up as mule loads and the barbed wire loads had gloves and pliers fastened to them.

12. *Field Hospitals.*—The field hospitals were No. 1 British and Nos. 101 and 102 Native. The furniture gave satisfaction, except the field hospital boxes which were supplied by the Supply and Transport Corps and Medical Store Depôt.

As the Barrack Department supplies identical boxes to troops on the line of march, it seems an unnecessary complication that the officer in charge of the field hospital should on mobilisation have to draw similar boxes from two sources, *viz.*, the Supply and Transport Corps and the Medical Store Depôt. I think it would simplify matters and be more satisfactory if all " Boxes Field Hospital " were supplied by the Barrack Department. " Type Plan No. 16, Box, Field Hospital " should be altered. The rings which attach the box to the saddle should be near the top and on the opposite side of the box to that shown in the plan. In the Peshawar District this alteration was introduced a few years ago.

13. *Employment of Royal Engineer Officers and Military Works Service Non-Commissioned Officers.*—No civil labour was employed. I therefore attached one officer (Lieutenant Turner, R.E.,) to the 23rd Sikh Pioneers and the remainder to the Sapper and Miner Companies.

One Barrack Sergeant (Sergeant Harvey), who had been through a Park Sergeant's course, was placed in charge of the advanced depôt of the Engineer Field Park and was responsible for the care and despatch of stores.

The officer in charge of No. 1 Engineer Field Park personally brought explosives to within one march of their destination.

14. *Organisation and equipment of Sapper and Miner Companies.*—I have no remarks to offer except that I think it should be an instruction to Pioneer regiments to occasionally test their detonators, primers and wet guncotton by exploding a few slabs, say, every six months. I also recommend that Brevet Major Sheppard's pamphlet on blowing down towers, etc., be again circulated to all officers of the Military Work Services.

15. *Steps taken preliminary to concentration.*—The steps taken before concentration were as follows :—

5th February 1908.—Tested and put right the Peshawar City station platform water-supply. Commenced packing up all stores for the advanced depôt at Ali Musjid.

6th February 1908.—Fitted up in the Military Work Service godown at Peshawar 3 water-supply standards of 26 taps each.

7th February 1908.—Commenced rigging up one of the above standards at Jamrud, this was connected with the main from Kacha Garhi and 200 feet length of troughing for animals. This troughing was filled by gravitation from the tank outside Jamrud fort.

12th February 1908.—Rigged up two of the above standards at Kacha Garhi, demolished toll bar and stone wall at Jamrud as they impeded movement. Sent to Jamrud barbed wire fencing and a Well's light for use in the Supply depôt. On this date I went to Jamrud to superintend the work.

13th February 1908.—The Field Force marched from Peshawar to Jamrud.

W. J. D. DUNDEE, *Lieut.-Col., R.E.,*
Commanding Royal Engineer, Bazar Valley Field Force.

12*th March* 1908.

APPENDIX 5.
Strength of Bazar Valley Field Force.

	British Officers.			British other rank	Native Officers.	Native other ranks.	Supply and transport personnel.	Public followers.	Private followers.	Officers' charges.	Horses, Native Cavalry.	Battery Ponies.	Pack ponies and mules, etc.	Riding ponies.	Camel.	Bullocks	REMARKS.
	British Service.	Indian Service.															
Marched out with General Sir J. Willcocks on 13th February 1908.	88	167*		1,712	159*	7,675*	2,503	1,047	552*	210	554	13	6,699	101	103	..	* These numbers include Khaibar Rifles.—7 British officers, 14 native officers, 702 Rank and File, and 29 private followers.
At Jamrud before force started.	..	4		17	..	1	101	132†	8	5	..	..	..	4	270	30	† Includes 119 men specially entertained.
Gone to Jamrud since 13th February 1908.	2	8		2	4	58	130	243‡	20	10	..	..	..	..	347	38	‡ Includes 228 men specially entertained.
Total west of Peshawar, 25th February 1908.	90	179		1,731	163	7,734	2,734	1,422	580	225	554	13	6,701	105	720	68	

In addition to the above, the following hired transport has been employed at Jamrud :—
175 camels, 1,282 donkeys, 165 carts, 6 hand-carts, 23 packbullocks and maundage for 1,389 maunds.
The hired transport is being dispensed with from 21st instant.

Now at Base	4	5		23	2	116	..	17	57	9	..	..	..	11	..	8	
GRAND TOTAL	94	184		1,754	165	7,850	2,734	1,439	637	234	554	13	6,701	116	720	76	

APPENDIX 6.

Report on signalling operations with Bazar Valley Field Force by Captain A. B. Whatman, Inspector of Army Signalling, Northern Circle.

The expedition came at an inopportune moment as far as signalling was concerned as a transition stage was taking place between two systems of visual signalling, namely :—The English method of transmitting messages, the Indian method of transmitting messages. The signallers of some units were sufficiently advanced in the new or English system, whilst others had not started to work on the new lines. This, taking into consideration that units had not received the English message form, necessitated reverting to the now obsolete Indian message form during the time the expedition lasted.

Five of sets signalling equipment were available in the Peshawar Fort. These were drawn and the personnel provided for by six signallers of the 80th Battery, Royal Field Artillery, fifteen signallers of the Northumberland Fusiliers, and fifteen signallers of the 57th Wilde's Rifles. Thus five special signalling units were established and were told off as follows :—

1. One unit to Divisional Head-Quarters.
2. ,, ,, ,, 1st Brigade.
3. ,, ,, ,, 2nd Brigade.
4. ,, ,, ,, Lines of Communications.
5. ,, ,, ,, ,, ,, ,,

A field telegraph was established at Fort MAUD and ALI MASJID. On February 15th on the advance of the Division into the BAZAR Valley a signalling station was established at ZERA formed by the 1st Brigade, the station being some distance below the summit of ZERA and immediately above CHURA Kotal. From this point communication was established with the Fort belonging to YAR MOHAMED in the village of CHURA. As the first day's (February 15th) objective of the 2nd Brigade was WALAI it was found necessary to establish a station on the ridge east of the village of BURG, as WALAI lies in a valley and the ridge to the east of BURG makes direct communication between ZERA and the village of WALAI impossible. On the evening of the 15th February communication was established between the first Brigade at CHURA and the second Brigade at WALAI through BURG.

On February, the 16th, it was found that from KHAR GHUNDAI a hill west of WALAI direct communication could be obtained with ZERA and CHURA, it was therefore no longer necessary to keep the station at BURG open and the signallers and escort (2 companies, 45th Sikhs) were withdrawn. On the morning of February 16th communication was established from KHAR GHUNDAI with Colonel Roos-Keppel's

Column in CHINA. From February 16th onwards communication was established from Head-Quarters (Camp WALAI east of CHINA) to Fort MAUD through KHAR GHUNDAI and ZERA. The hill KHAR GHUNDAI is the key to signalling operations in the BAZAR Valley both as regards lines of communication and for any forward operations which may take place west of the village of WALAI. On the 19th February the special signalling unit was withdrawn from Fort JAMRUD and sent forward to supplement the post at ZERA. Two acetylene lamps were obtained from KASAULI for work between ZERA and KHAR GHUNDAI, the mist at night making work at times impossible between these stations with B. B. Lamps.

The fact of there being no field telegraph beyond FORT MAUD made the signalling at times heavy, but the signallers responded gamely and put in some very sound work. Some of the messages were of considerable length, ranging from 500 to 900 words, including cipher and figures which necessitates repetitions and so prolong the length of a message. A prominent and satisfactory feature was the accuracy of the work, no time was wasted and the results only tend to shew the very thorough and careful training which is given by regimental signalling officers. It should also be noted that the great majority of signallers employed during the expedition belonged to the Indian Army. I attribute the efficiency to the following facts :—

1. The great interest which is taken in the training of signallers in the great majority of units.
2. That during the expedition fast and inaccurate sending was not allowed, and that a uniform rate of eight words a minute was enforced.
3. The presence of a British Officer at the main signalling stations on the lines of communication.

The following officers were employed as signallers (not including those signalling officers with their regiments).

1st Brigade	Major Ferguson-Davie, C.I.E., D.S.O., 53rd Sikhs.
2nd ,,	Lieutenant McLeod, Guides Cavalry.
FORT MAUDE	Lieutenant Gibson, Northumberland Fusiliers.
ZERA	Lieutenant Barrow, 38th Dogras. * Lieutenant Sandeman, Guides Infantry.
BURG	Lieutenant Ramsford-Hannay, 45th Sikhs.

* At different periods.

KHAR GHUNDAI . . . Lieutenant Burrows, 28th Punjabis. ⎫
Lieutenant Ramsford-Hannay, 45th Sikhs. ⎬ *
Lieutenant Barrow, 38th Dogras. ⎭

An improvement has taken place, in that signallers take more interest in their work, and consequently are more on the *qui vive* to take and receive a message in the field. There is, however, great room for improvement in this respect and some signallers still walk about with their eye on the ground. A signaller should constantly be on the lookout and every signalling officer should impress on his men the paramount importance of this duty in the field; and for this purpose a signaller must always consider himself on duty.

A. B. WHATMAN, *Captain,*
Inspector of Army Signalling, Northern Circle.

---

* At different periods.

APPENDIX 7.

Statement of ammunition expended by the Zakka Khel (Bazar Valley) Field Force from 13th to 29th February 1908.

Corps.	Shrapnel shell.	Common shell.	Total shells.	305" rifle.	303" maxim.	303" Carbine.	M. H.
1st Royal Warwicks	..	..	..	5,530			
23rd Pioneers	..	..	..	3,939			
53rd Sikhs	..	..	..	14,708	562		
59th Rifles	..	..	..	15,585	880		
2-5th Gurkhas	..	..	..	11,540	,092		
22nd Derajat Mountain Battery.	564	·7	571	..	..	35	
3rd Mountain Battery, Royal Garrison Artillery.	462	1	463				
Seaforths	..	..	..	20,136			
20th Punjabis	..	..	..	6,504			
45th Sikhs	..	..	..	2,781	557		
54th Sikhs, Frontier Force	..	..	..	16,473			
No. 6 Company, Sappers and Miners	..	..	..	365			
37th Lancers	..	..	..	3			
19th Lancers	..	..	..	32			
Khaibar Rifles	..	..	..	..	..	..	7,885
			1,034	100,687		35	7,885

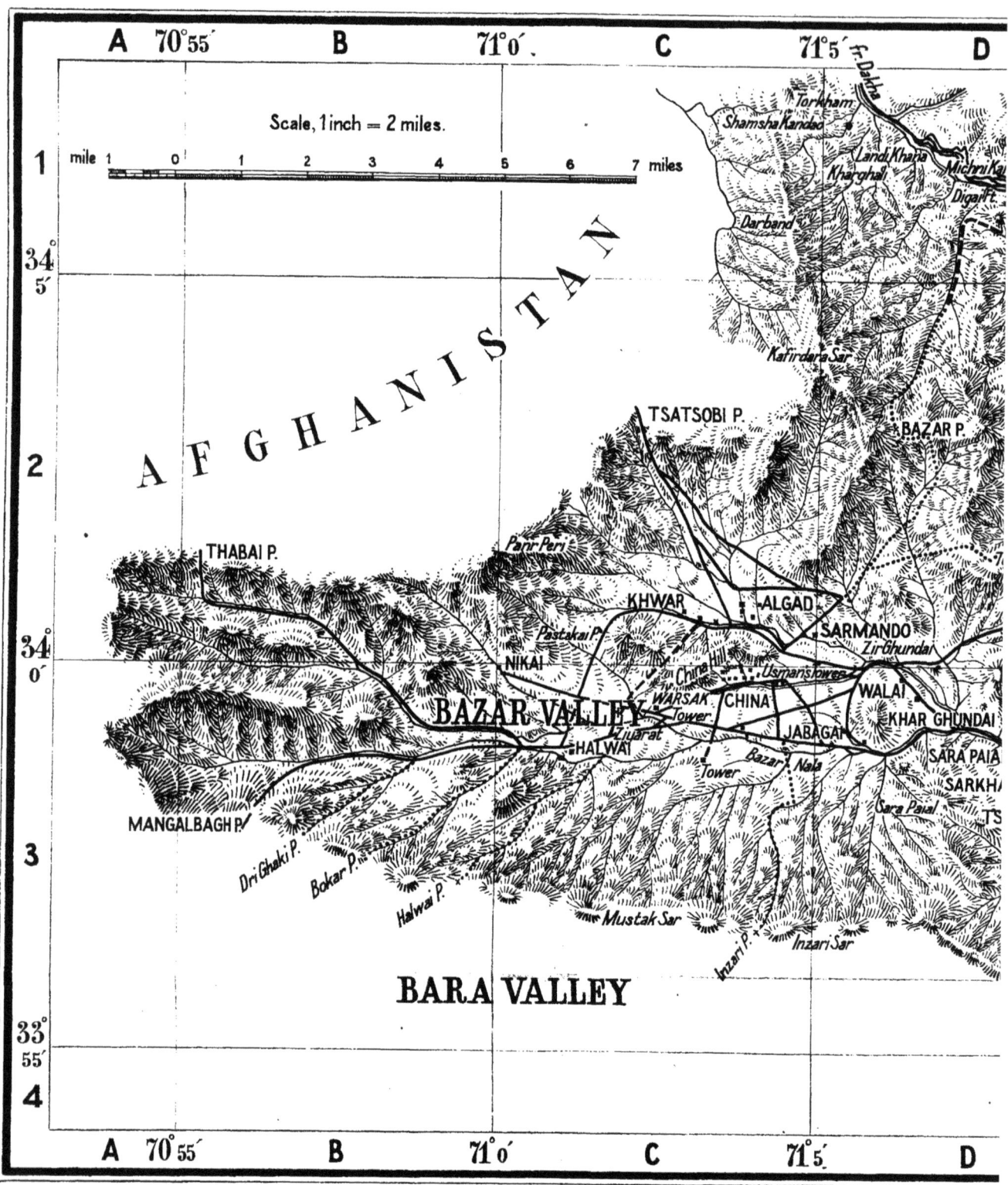

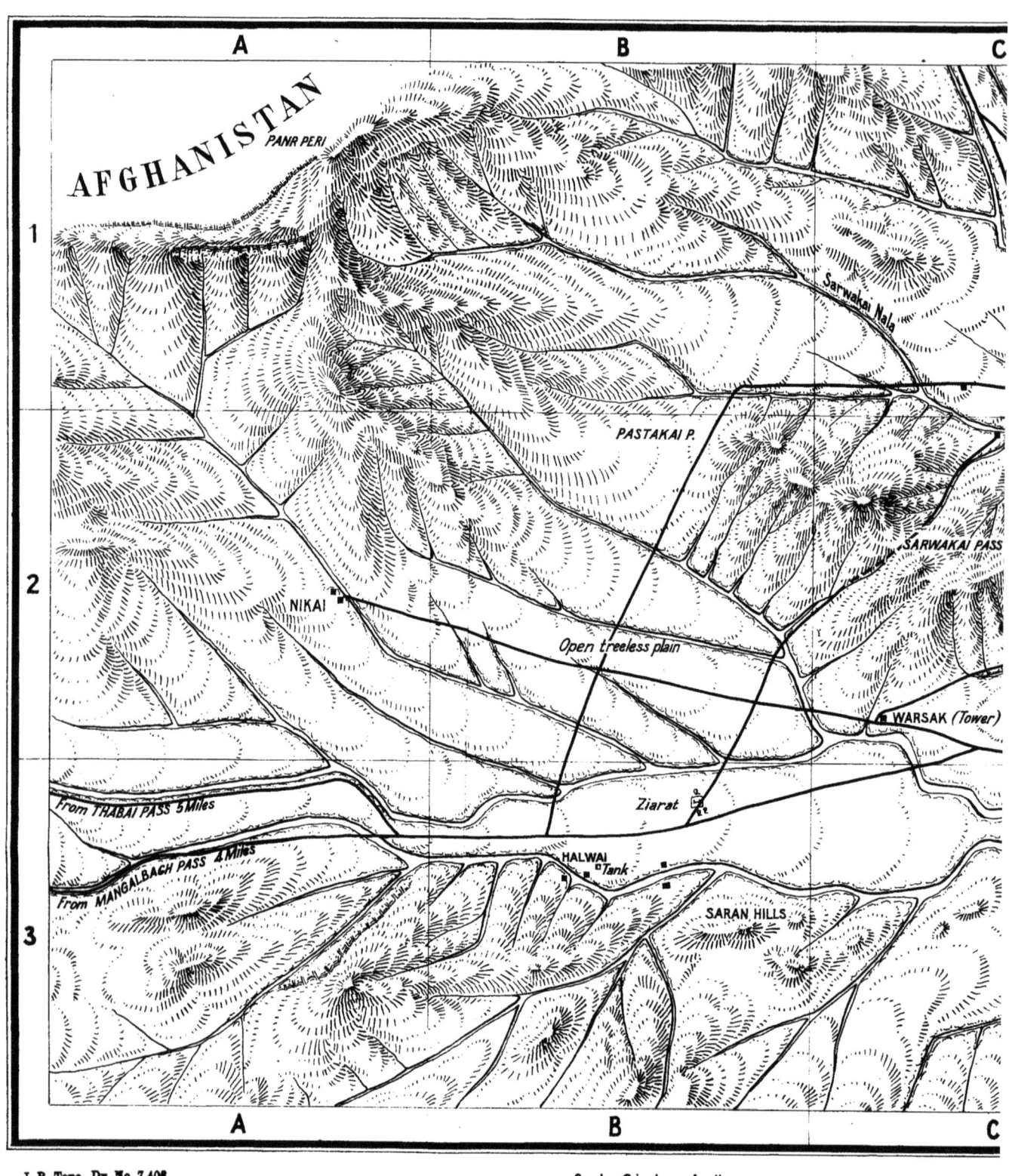

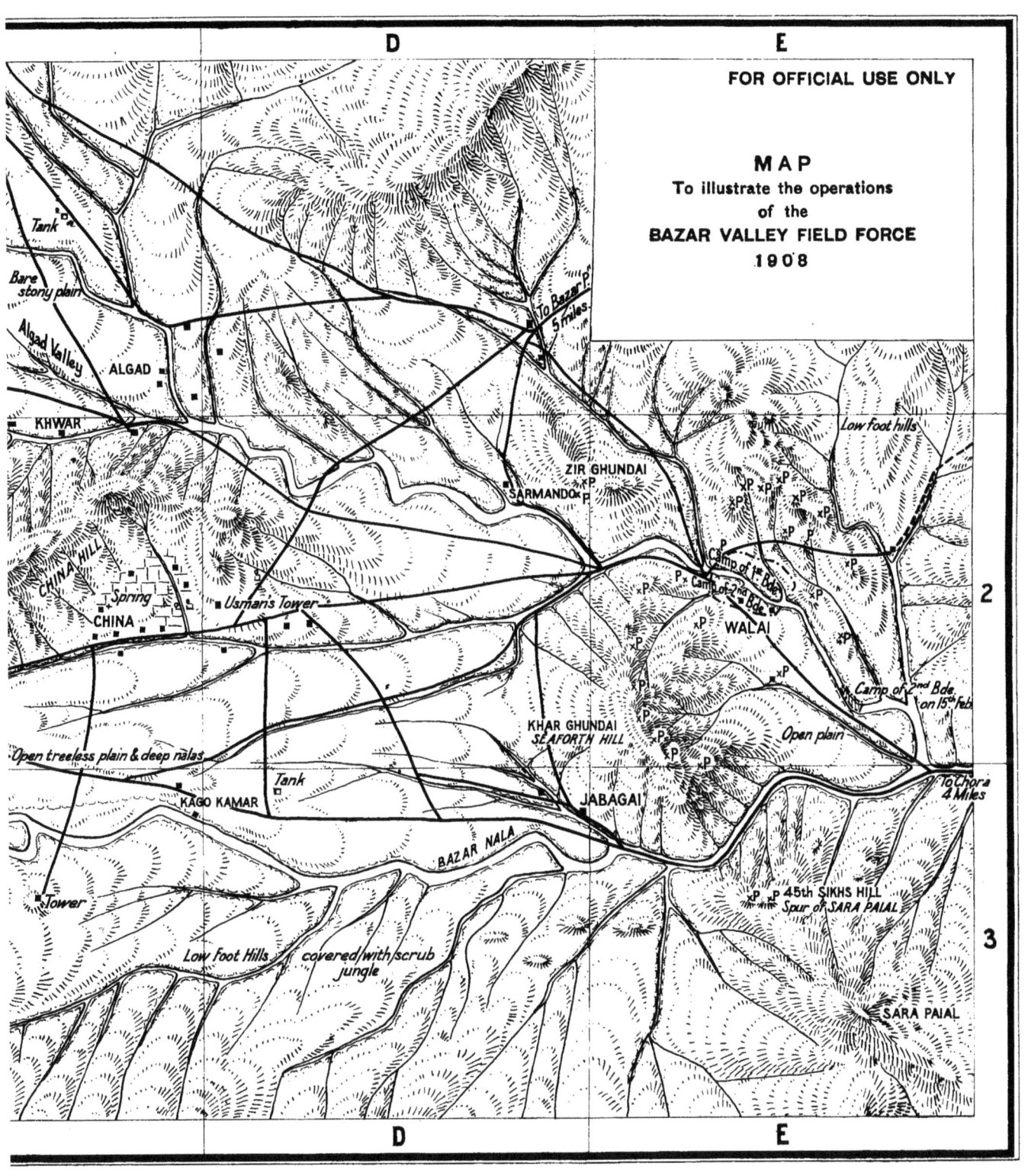

MAP to illustrate SIGNALLING OPERATIONS, BAZAR VALLEY
February 1908.

Scale 1 inch = 2 miles.

REFERENCE

Communications established ————
do.   which are possible - - - - -

(Sd.) A. B. Whatman, Capt.,
Inspector Army Signalling,
9.3.08.   N.C

No. 4,525.,-I., 1908.

I. B. Topo. Dy. No. 7,404.
Exd. C. J. A. August 1908.

www.ingramcontent.com/pod-product-compliance
Ingram Content Group UK Ltd.
Pitfield, Milton Keynes, MK11 3LW, UK
UKHW050416240426
12048UKWH00021B/1546